Giacomo Bertini

Sampler systems for tracking emitters of phenanthrene in sewers

AF011097

Giacomo Bertini

Sampler systems for tracking emitters of phenanthrene in sewers

Südwestdeutscher Verlag für Hochschulschriften

Impressum / Imprint
Bibliografische Information der Deutschen Nationalbibliothek: Die Deutsche Nationalbibliothek verzeichnet diese Publikation in der Deutschen Nationalbibliografie; detaillierte bibliografische Daten sind im Internet über http://dnb.d-nb.de abrufbar.
Alle in diesem Buch genannten Marken und Produktnamen unterliegen warenzeichen-, marken- oder patentrechtlichem Schutz bzw. sind Warenzeichen oder eingetragene Warenzeichen der jeweiligen Inhaber. Die Wiedergabe von Marken, Produktnamen, Gebrauchsnamen, Handelsnamen, Warenbezeichnungen u.s.w. in diesem Werk berechtigt auch ohne besondere Kennzeichnung nicht zu der Annahme, dass solche Namen im Sinne der Warenzeichen- und Markenschutzgesetzgebung als frei zu betrachten wären und daher von jedermann benutzt werden dürften.

Bibliographic information published by the Deutsche Nationalbibliothek: The Deutsche Nationalbibliothek lists this publication in the Deutsche Nationalbibliografie; detailed bibliographic data are available in the Internet at http://dnb.d-nb.de.
Any brand names and product names mentioned in this book are subject to trademark, brand or patent protection and are trademarks or registered trademarks of their respective holders. The use of brand names, product names, common names, trade names, product descriptions etc. even without a particular marking in this work is in no way to be construed to mean that such names may be regarded as unrestricted in respect of trademark and brand protection legislation and could thus be used by anyone.

Coverbild / Cover image: www.ingimage.com

Verlag / Publisher:
Südwestdeutscher Verlag für Hochschulschriften
ist ein Imprint der / is a trademark of
OmniScriptum GmbH & Co. KG
Heinrich-Böcking-Str. 6-8, 66121 Saarbrücken, Deutschland / Germany
Email: info@svh-verlag.de

Herstellung: siehe letzte Seite /
Printed at: see last page
ISBN: 978-3-8381-3517-5

Zugl. / Approved by: Essen, UDE, Diss., 2014

Copyright © 2015 OmniScriptum GmbH & Co. KG
Alle Rechte vorbehalten. / All rights reserved. Saarbrücken 2015

TABLE OF CONTENTS

ABSTRACT — 3

1 INTRODUCTION — 5

1.1 Pollution in sewage sludge and wastewater — 5
1.1.2 Priority substances, polycyclic aromatic hydrocarbons (PAHs) and origin of contamination — 9
1.1.2.1 Regulation — 9
1.1.2.2 Occurrence of PAH pollutants — 15
1.1.2.3 Properties of PAHs and phenanthrene — 20
1.1.2.4 Octanol-water partition coefficient (Kow) — 28
1.1.2.5 Analytical methods to detect phenantherene — 30

1.2 Microbial biofilms — 34

1.3 Fate of PAHs in Sewers and role of sewer biofilms — 36
1.3.1 Sewer biofilms — 38
1.3.2 Sorption properties of Biofilms — 40

1.4 Biodegradation of polycyclic aromatic hydrocarbons — 46

1.5 Use of the sorption memory of biofilms and polydimethylsiloxane (PDMS) in order to locate pollutants in sewers. — 51

1.6 Polysaccharide gels as samplers for PAH pollutants in water systems. — 55
1.6.1 Polysaccharide gels and microbial biofilms: sorption of hydrophobic pollutants — 56

1.7 Aims of this study — 63

2 MATERIALS AND METHODS — 64

2.1 Materials — 64
2.1.1 Polydimethylsiloxane (PDMS) monitor device — 64
2.1.2 Optical fiber platform — 65

2.2 Analytical Instruments — 66
2.2.1 Bench fluorescence spectrometer — 66
2.2.2 USB fluorescence spectrograph — 66
2.2.4 GC-MS device — 68

2.3 Methods — 68

Table of contents

    2.3.1 Preparation of aqueous solutions of phenanthrene _____ 68
        2.3.1.1 PDMS device experiment in deionized water solution _____ 68
        2.3.1.2 PDMS device experiment in pond water solution _____ 69

2.4 Sampling of the sewer biofilm _____ 70

2.5 Front face fluorescence measurements on the PDMS device _____ 72

2.6 Calibration procedures _____ 73
    2.6.1 The quantification of phenanthrene in deionized water solution. _____ 73
    2.6.2 GC-MS device _____ 74

2.7 Experiments _____ 75
    2.7.1 PDMS device in deionized water with phenanthrene _____ 75
    2.7.2 PDMS device in pond water with phenanthrene _____ 75

2.8 Data elaboration procedures _____ 76
    2.8.1 Simulating diffusion of phenanthrene from the self-designed PDMS device. _____ 76

3 RESULTS _____ 77
3.1 Quantification of phenanthrene in sewer biofilms by GC-MS _____ 77

3.2 PDMS Device as pollutant sampler and analytical device _____ 78

4 DISCUSSION _____ 91
4.1 Sewer biofilms as memory of PAHs in sewers _____ 91

4.2 PDMS device for detection of phenanthrene in sewers _____ 94

5 REFERENCES _____ 102

6 APPENDIX _____ 126
6.1 List of abbreviations _____ 126

ACKNOWLEDGEMENTS _____ 129

ABSTRACT

This book refers to part of a doctoral dissertation, which was written at the University of Duisburg-Essen. It contains data from a review study that was carried out within a Marie Curie FP7 initiative called ATWARM (Advanced Technologies for Water Resource Management).

Pollutants can accumulate in sewage sludge. In such cases, this sludge represents toxic waste and has to be disposed on specialized dumps or eliminated by incineration. The mechanisms of sorption and the sorption sites for such substances in sludge are varying according to their physico-chemical properties. Fact is that both polar and non-polar substances can be accumulated. The development of a monitoring approach for detecting pollutants along sewers represents a suitable solution for locating polluters upstream the waste water treatment plant (WWTP) and prevent the contamination of the sewage sludge.

First, a long review about the environmental occurrence, emission and properties of polycyclic aromatic hydrocarbons (PAH) is presented. Phenanthrene is one of the most common PAH pollutants found in sewage sludge and was chosen in this work as a reference compound.

It was reported that microbial biofilms can absorb and accumulate pollutants in aqueous environments, therefore sewer biofilms can be considered as sampler system for detecting PAHs along the sewers. In addition to their role as a sink for pollutants, biofilms can desorb the absorbed compounds back into the water. The period of time when the absorbed pollutant is detectable inside the biofilm is defined as the memory. It is necessary to estimate the memory of biofilms for phenanthrene in order to develop a systematic approach for monitoring PAH compounds in sewers using these specific matrices.

Abstract

The manuscript is focused on the use of sewer biofilms, polysaccharide gels and polydimethylsiloxane (PDMS) as sorbent materials in passive sampler devices for monitoring PAHs in wastewater and surface water streams.

A short experimental section is also presented in the book. The results of this part contributed to shed more light on the topic and are involved in the final discussion, where the potential of sewer biofilms and PDMS oil for monitoring PAH in wastewater is evaluated. Phenanthrene was not detected in the biofilm samples taken from the monitored wastewater system. On the other hand, PDMS oil exhibited a high potential for developing passive sampler devices for the detection of phenanthrene in water by on field measurements (fluorescence spectroscopy). In this regard, a novel monitor device was designed here and proposed for future applications.

During all the essay, particular attention is addressed on the memory effect of sewer biofilms for hydrophobic pollutants. What is known about the memory and about the desorption kinetics of hydrophobic pollutants from biofilms in flowing water systems is presented by reviewing the most significant studies reported within scientific literature. Also a notional comparison between the diffusion properties of polysaccharide gels and microbial biofilms is implemented in the text and it well conforms to the content of the whole investigation. It is explained in this part how limited is the ability of the most commonly used polysaccharide gels for simulating the sorption properties of biofilms.

Although only little has been investigated on this topic, the memory effect for absorbed pollutants in microbial biofilms is the crucial parameter which describes their potential as sampler systems for environmental moitoring purposes. More research should be addressed on this particular aspect.

1 INTRODUCTION

1.1 Pollution in sewage sludge and wastewater

Before the establishment of the modern wastewater treatment systems, most of the communities discharged their wastewater, or sewage, into streams and rivers with little if any treatment. As urban populations increased, the quality of streams and rivers began to deteriorate in many regions. In response to the water quality degradation and to concerns about healthcare issues, the need of suitable wastewater treatment systems arose.

"By the end of the 18th century, all major northern European cities had built, or were building new systems to distribute water and evacuate liquid wastes (Reid, 1991). The realization that many diseases, such as cholera, were passed on by contaminated water was primarily behind the development of this "big pipe engineering approach" - for bringing drinking water into the city and for removing wastewater an d storm water from the city. Big pipe engineering became the standard water management technique" (Vigneswaran et al., 2009)

The foundation of wastewater treatment plants and the establishment of waste water management strategies greatly improved stream and river water quality of the modern communities, but created another material to deal with: sewage sludge. The wastewater that enters a treatment plant is released from a variety of sources including homes, industries, medical facilities, rural activities, street runoff or businesses. Most of the water volume turns into clean effluent by the treatment processes, while the remainder portion is a dilute suspension of solids that has been captured by the primary, secondary and advanced treatment processes and includes grit, screenings and sludge. Of these constituents, sludge is by far the largest in volume, therefore handling methods and disposal practices of it are a matter of great concern. Since 1988, when the European legislation phased out the seawater disposal of the sludge (by the Urban Waste Water Treatment Directive), the main practices for handling the sewage sludge are: agricultural use (37%), incineration (11%), landfilling (40%), forest sulviculture and land reclamation (12%).

The latest trends in the field of sludge management are: combustion, wet oxidation, pyrolysis, gasification and co-combustion of sewage sludge with other materials. All of these new handling strategies have the aim to use the sludge as energy source. As previously mentioned the recycling of the waste sludge to land fertilization and land filling are the most common practices (Fytili and Zabaniotou, 2008) and it is important to lower their contamination levels as much as possible in order to avoid any further environmental pollution. The pollutants present in the sludge can be divided in three groups:

1. Potential toxic elements (heavy metals)
2. Organic pollutants
3. Pathogens.

The fate of chemical contaminants entering a waste water treatment plant (WWTP) depends on both the nature of the chemical and the treatment processes (Zitomer and Speece, 1993). Because of the physical–chemical processes that are involved in activated wastewater sludge treatment, sludge tends to accumulate heavy metals existing in the wastewater. Heavy metals such as zinc (Zn), copper (Cu), nickel (Ni), cadmium (Cd), lead (Pb), mercury (Hg) and chromium (Cr) are principal elements restricting the use of sludge for agricultural purposes (US EPA 1993; The Sewage Sludge Directive 86/278/EEC; Hsiau and Lo; 1998).

The heavy metals are potential toxic elements (PTE) for human health and show long-term accumulation in soils and sediments (Wuana and Okieimen; 2011). The majority of PTEs entering the wastewater treatment plant transfer to the sewage sludge but between 20% and 40% can be released in the effluents from the treated wastewater. According to the EC policy of waste recycling, recovery and use, the application of sludge to agricultural land must be coupled with a constant monitoring of its contamination levels. This is a critical issue in view of the fact that the amount of sludge produced is going to increase and the governmental policies are going to be

1 Introduction

more stringent. Hence sludge quality must be protected and improved in order to secure the agricultural outlet as the most cost effective and sustainable option. The main source of PTE are the commercial and industrial activities. In Table 1 are shown the main inorganic pollutants occurring in sewage sludge along three European areas and the percentage of total incoming for each main source.

Table 1 Fytili and Zabaniotou; 2008 Contribution to the total PTE pollution of different sources, in reference areas in Europe (% values).

Pollutant	Country	Domestic wastewater	Commercial wastewater	Urban Runoff	Not identified
Cd	France	20	61	3	16
	Norway	40			
	UK	30	29	41	
Cu	France	62	3	6	29
	Norway	30			
	UK	75	21	4	
Cr	France	2	35	2	61
	Norway	20			
	UK	18	60	22	
Hg	France	4	58	1	37
Pb	France	26	2	29	43
	Norway	80			
	UK	43	24	33	
Ni	France	17	27	9	47
	Norway	10			
	UK	50	34	16	
Zn	France	28	5	10	57
	Norway	50			
	UK	49	35	16	

Organic chemicals may be volatilized, degraded (through biotic and/or abiotic processes), sorbed to sludge or discharged in the aqueous effluent. The main category of organic pollutants are: polycyclic aromatic hydrocarbons (PAHs), Polychlorinated biphenyls (PCB), di-2-ethylhexyl phthalate (DEHP) and Polychlorinated dibenzodioxins/dibenzofuran (PCDD/PCDF). These compounds are relatively hydrophobic and are strongly retained into particles and sludge. Most of them are not easily biodegraded during the treatments and can represent an environmental threat. Even though some of them are biodegraded, the by-products might be harmful and dangerous for the environment (European commission, *"Pollutants in waste water and sewage sludge"*, 2014, web). The main reason why these pollutants are a major concern is their harmful effect on human beings and other living organisms. Both the

1 Introduction

European community and the US environmental protection agency (EPA) established limit values of concentration for these pollutants within their own waste water framework policies and review these directives periodically in order to guarantee a suitable regulation according to the growing levels of industrialization and urbanization.

Since the time of the directive 86/278/EEC more pathogens associated with the food chain have been identified and new technologies have become available for sludge treatment. After separation from the wastewater, the sludge must be treated through one of a number of processes. Each of these has effects on the fate of both pathogens and the organic contaminants in the sludge (Rogers, 1996). Heavy metals are not removed throw these steps and therefore their concentration must be evaluated before to address the sludge to land recycling purposes.

In order to produce sludge that is considered free of pathogens, it is exposed to thermophilic biological treatment and microbial digestion; or a combination of high pH and relatively high temperature (Table 2).

Table 2 Overview of the most used treatment processes of sewage sludge (European commission report; 2001).

Process	Parameters
Windrow composting: production of compost by piling biodegradable waste in long rows. This method is suited to producing large volumes of compost.	Batches of sludge (+/- bulking agent) to be kept at 55°C for 4 hours between each of 3 turnings, followed by maturation period to complete the composting process.
Aereted pile and invessel composting	The batch to be kept at a minimum of 40°C for at least 5 days and for 4 hours during this period at a minimum of 55°C. This to be followed by maturation period to complete the composting process.
Thermal drying	The sludge should be heated to at least 80°C for 10 minutes and moisture content reduced to < 10%.
Thermophilic digestion (aerobic or anaerobic)	Sludge should achieve a temperature of at least 55°C for a minimum period of 4 hours after the last feed and before the next withdrawal. Plant should be designed to operate at a temperature of at least 55°C with a mean retention period sufficient to stabilise the sludge.
Heat treatment followed by digestion	Minimum of 30 minutes at 70°C followed immediately by mesophilic anaerobic digestion at 35°C with a mean retention time of 12 days
Treatment with lime (CaO)	The sludge and lime should be thoroughly mixed to achieve a pH value of at least 12 and a minimum temperature of 55°C for 2 hours after mixing.

Concerning all the management strategies, which are used to handle the sewage sludge, pros and contras must be always considered carefully. Therefore, it is still a priority to monitor the waste water distribution systems and set suitable prevention strategies in order to lower the contamination at the source level.

1.1.2 Priority substances, polycyclic aromatic hydrocarbons (PAHs) and origin of contamination

1.1.2.1 Regulation

The priority substances are contaminants, which pose a significant risk to or via the aquatic environment. These substances are selected amongst those presenting higher index scores based on the occurrence in the environment and the toxicity on living organisms and also on human health (Lerche et al.,2002). Polycyclic aromatic

1 Introduction

hydrocarbons (PAHs) represent a category of highly common pollutants detected in water, soils, atmospheres and sewage sludge (Harrison et al., 2006, Perez et al., 2001). Some of them present an extremely low solubility in water and these are mostly accumulated in soil and living organisms. In addition, they display harmful effects on human beings, such as toxicity, mutagenicity and carcinogenicity.

The ecotoxicology of PAHs can be investigated by different tests involving the use of reference organisms like microorganisms, small invertebrates and plants (Table 3).

Table 3 Ecotoxicology tests of PAHs on contaminated soil samples (Eom et al.,2007).

	Tests (species)	Duration	Measured parameters	Test procedure
Microorganisms	Microtox® (*Vibrio fischeri*)	15, 30 min	Luminescence inhibition	ISO NF EN 11348-3 (1999)
	Mutatox® (*Vibrio fischeri* M169)	24 h	Genotoxicity	Microbics (1993)
	Ames (*Salmonella typhimurium* TA98, TA100)	72 h	Genotoxicity	Environment Canada (1993)
	Umu (*Salmonella typhimurium* TA1535/pSK1002)	4 h	Genotoxicity	ISO/DIS 13829 (2000)
Algae	Algae (*Pseudokirchneriella subcapitata*)	72 h	Growth inhibition	ISO 8692 (1996)
Aquatic invertebrates	Daphnid (*Daphnia magna*) Ceriodaphnid (*Ceriodaphnia dubia*)	24, 48 h 7 days	Immobilization reproduction	ISO 6341 (1996), AFNOR T90-376 (2000)
Terrestrial invertebrates	Collembola (*Folsomia candida*)	28 days	Survival, reproduction	ISO 11267 (1999)
	Earthworm (*Eisenia fetida*)	14 days 28, 56 days	Survival, reproduction	ISO 11268-1 (1993), ISO 11268-2 (1998)
Terrestrial plants	Lettuce (*Lactuca sativa*) Chinese cabbage (*Brassica chinensis*)	14–21 days	Germination, growth	ISO 11269-2 (1995)

Despite their high toxicity on most aquatic and terrestrial organisms, the priority in performing risk assessment investigations on contaminated soils and water streams is to evaluate the toxicity on mammals in reference to human beings. The toxicity of PAHs on mammals has been largely studied by testing mice and rats in laboratory conditions and evaluating the LD50 (letal dose of 50% of the tested subjects) and other effects such as nephrotoxicity, oocyte and follicle destruction, testicular damage, carcinogenicity. In this regards Eisler 1987 provides a review that

1 Introduction

summarizes interesting results about the toxicity (Table 4) of both carcinogenic (Figure 1) and non-carcinogenic (Figure 2) PAHs.

Benz(a)anthracene (BaA) $C_{18}H_{12}$

Crysene (Chr) $C_{18}H_{12}$

Benz(a)pyrene (BaP) $C_{20}H_{12}$

Dienz(a,h)anthracene (DBahA) $C_{22}H_{14}$

Benz(b)fluorenthene (BbF) $C_{20}H_{12}$

Indeno(1,2,3-c,d)pyrene (IPyr) $C_{22}H_{12}$

Benz(k)fluoranthene (BkF) $C_{20}H_{12}$

Naphthalene (Nap) $C_{10}H_8$
Classified only as potential carcinogenic by the US EPA.

Figure 1 carcinogenic PAHs listed by the US EPA and by the European community. The name, the abbreviation and the structure are showed for each of the eight most regulated ones (Jennings et al.,2012).

1 Introduction

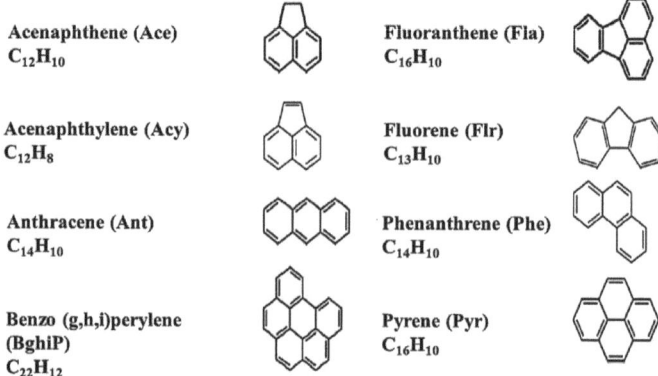

Figure 2 Non-carcinogenic PAHs listed by the US EPA and by the European community. The name, the abbreviation and the structure are showed for each of the eight most regulated ones (Jennings et al.,2012).

Table 4 Results of toxicity tests on rodents. In the right column are the values of PAH limit concentration for each toxicological test presented on the left column.

Tests

LD50 on rodents	mg/kg of body weight
Benzo-a-pyrene	50
Phenanthrene	700
Naphtalene	1780
Fluoranthene	2000
Carcinogenity oral administered	mg/kg of body weight
Benzo-a-pyrene	0.002
Chrysene	99
Anthracene	3300
Carcinogenity, topical administered	mg
Benzo-a-pyrene	0.001
Anthracene	0.08
Testicular damage oral administered	mg
Benzo-a-pyrene	100
Nephrotoxicity single intraperitoneal injection	mg/kg of body weight
Phenanthrene	150
Pyrene	150

1 Introduction

In Europe the first policy notice about priority substances was included in the article 16 of the Water Framework Directive (2000/60/EC, WFD). This directive sets out "Strategies against pollution of water", outlining the need to establish, by way of Decision 2455/2001/EC, a first list of priority substances to become Annex X of the WFD. Later a more specific definition of priority substances was provided by the Directive on Environmental Quality Standards (Directive 2008/105/EC, EQS directive), which establishes environmental quality standards for the substances in surface waters (river, lake, transitional and coastal) and confirmed their designation as priority or priority hazardous substances. According to Annex V, point 1.4.3 of the WFD and Article 1 of the environmental quality standards directive, good chemical status is reached for a water body when it complies with the environmental quality standards for all the priority substances and other pollutants listed in Annex I of the environmental quality standard directive (European commission, web page, Feb. 2014).

Each of the PAHs mentioned above, has been deeply studied and used as marker for environmental monitoring approaches. Although the establishment of reference environmental quality standards and the implementation of environmental guidelines by the European community and the US EPA, most of the regulation is still subject to country-specific laws.

For surface soil, the regulatory guidance values around the world can greatly vary from country to country. Jennings et al., 2012 provides a brief summary of these differences (Table 5). The Benzo(a)pyrene (BaP) is the reference compound for carcinogenic PAHs and the total concentration of the eight PAHs from Fig.2 is often used as reference for the limit concentration of non-carcinogenic PAHs.

Table 5 Limit values for carcinogenic (mg/Kg) PAHs in some of the countries mentioned by Jennings et al., 2012. The values which are not available are indicated with "-".

Country	Limit values (mg/kg)						
	BaA	BaP	BbF	BkF	Chr	DBahA	IPyr
US EPA	0.15	0.015	0.15	1.5	15	0.015	0.15
Germany	-	2	-	-	-	-	-
Belgium	10.5	3.6	7	11.5	180	2.9	20
Italy	0.5	0.1	0.5	0.5	5	0.1	0.1
Brazil	20	1.5	-	-	-	0.6	25

Also in regard to the non-carcinogenic PAHs, the regulatory guidance values are provided within a broader range of higher concentrations (Table 6).

Table 6 Limit values for non-carcinogenic PAHs in some of the countries mentioned by Jennings et al., 2012. T is the total PAH concentration. The values which are not available are indicated with "-".

Country	Limit values (mg/kg)							
	Ace	Acy	Ant	BghiP	Fla	Flr	Phe	Pyr
US EPA	3400	-	17000	-	2300	2300	-	1700
Germany	36-T	36-T	36-T	36-T	36-T	36-T	36-T	36-T
Belgium	14	1	70	3920	30	3950	65	395
Italy	-	-	5	0.2	5	-	5	5
Brazil	-	-	-	-	-	-	40	-

The regulations for PAHs levels in surface and drinking water are more uniform and shared at international level. In 1984 the World Health Organization (WHO) recommended values of total PAH concentration around 0.7 µg/L as risk limit for human health. Hennion et al., 1994 reports values of 0.2-1 µg/L at European level, while later in time, within the WFD 2000/60/EC the value was lowered to 0.1 µg/L. The US EPA sets the limit of 0.2 µg/L. However, different specific caveats in the regulatory system of each country must be considered also for these values.

At European level the WFD 2000/60/EC the limit value for total PAH concentration in sewer sludge is equal to 6 mg/kg, after a deep investigation on the real levels of these compounds in various urban waste water treatment plants around Europe, ranging from 0.5 mg/kg to 27.8 mg/kg and updated by 2014.

1.1.2.2 Occurrence of PAH pollutants

PAH compounds are a major concern of pollution due to their toxicity and their ubiquitous occurrence in air, soil, surface water and waste water, from which they accumulate in the sewage sludge (Srogi; 2007, Baek et al.,1991).

On a global scale around 85% of the emitted PAHs is represented by low molecular weight PAHs, such as naphtalene, phenanthrene, acenaphtilene, acenaphthene, fluorine, fluoranthene, pyrene, anthracene, while the high molecular weight PAHs cover only the least of the overall emission (Table 7, Zhang and Tao; 2009).

Table 7 Percentage of each low molecular weight PAH in the whole global emission (Zhang and Tao 2009)

PAH	% global emission
Naphtalene	50
Phenanthrene	10
Acenaphtylene	10
Acenaphtene	5
Fluorine	3
Fluoranthene	3
Pyrene	4

The PAHs sources can be both natural and anthropogenic. PAHs are mostly formed during the incomplete combustion and pyrolysis of fossil fuels or wood, and from the release of petroleum products. Other sources are petroleum spills, oil seepage and diagenesis of organic matter in anoxic sediments. The main sources are divided in five categories: domestic, mobile, industrial, agricultural and natural (Ravindra et al., 2008). Burning of coal, oil, gas, garbage and, to a lesser extent, other organic substances like food (e.g. char broiled meat) are the main processes belonging to the domestic source of PAH pollution. These processes are involved mostly in heating and cooking activities. The WHO estimated that 75% of the people in China, India and South east Asia burn solid fuels like wood, animal dung cakes and crop waste for daily heating and cooking. In a global scale these and the biofuel combustion are significant contributors to the overall emission of all the 16 PAHs listed by US EPA and European community as priority substances (Zhang and Tao, 2009, Table 8 and 9).

Table 8 The principal global emission processes of the 16 PAHs listed by the US EPA and the European community (Zhang and Tao; 2009).

Processes	Tons/year global emission
Open fires burning of wheat	10000
Indoor burning of corn	5000 – 10000
Indoor burning of rice	10000
Indoor burning of wheat	60000 – 80000
Forest fires	40000
Firewood burning	16000
Domestic combustion of coal	20000
Small scale coke production	10000 – 20000
Traffic gasoline	20000
Consumer products usage	20000-30000
Animal dung combustion	20000 – 30000
Grassland fires	20000 – 30000
Waste incineration	5000 - 10000

Table 9 Sources of PAHs. On the right is their percentage contribution on the global emission of the 16 PAHs listed by US EPA and European community as priority substances (Zhang and Tao 2009).

Sources of PAHs	% contribution
Biofuel burning	56.7
Wildfires	17
Consumer products	6.9
Traffic oil combustion	4.8
Domestic coal combustion	3.7
Industrial activities	10 (3.6% coke production)

# 1 Introduction

The Environment Directorate-general of the European Community quantified the total emission of benzo(a)pyrene in Europe in 1990. The report for the economic evaluation of the air quality targets for PAHs published in 2001 shows that also in Europe, between 1990 and 2010, the main sourcesof benzo (a)pyrene were those involving domestic and small burning processes of coal and wood (about 200 Tons/year), while the industrial activities are just minor contributors (between 2 and 50 tons/year). Furthermore between 1990 and 2010 the European commission aimed a significant reduction of the industrial emission of B(a)P, but a relatively small decrease in the wood burning processes, mainly due to a not significant increase in the use of other bio fuels rather than wood, which still represent a significant source of B(a)P.

However the relative contribution of different PAH sources in the different countries depends on the energy structure, status of development, population density and vegetation cover of the country, so that big differences are observed for specific PAH sources between the European area, USA and the east of the world (Zhang and Tao, 2009).

Another important group of sources is represented by the mobile sources. These ones are considered all the transport systems, which involve the consumption of diesel or gasoline. Aircrafts, automobiles, ships, railway locomotives, off road vehicles and machinery belong to this category. Among these the ones which are fueled by diesel are the stronger PAH producers. The main industrial sources are aluminum production, coke production, waste incineration, cement manufacture, petrochemical industries, bitumen and asphalt industries and rubber tire manufacturing. In addition, agricultural activities are considered one of the main sources of PAHs in the environment, due to the open burning of biomass that is employed for residue disposal and land preparation. All these procedures involve burning under sub optimum combustion conditions; therefore they are expected to significantly contribute to the overall PAH emission in the environment. The less important source of PAHs are natural processes such as forest and wood land fires (e.g. after lightning strikes) and volcano activities. Others are high temperature pyrolysis of organic matter, diagenesis of sedimentary organic material for fossil fuel formation and microbial biosynthesis.

As mentioned earlier PAHs are ubiquitous pollutants in the environment and cause a growing concern due to their significant harmful potential on human health. Toxicity, mutagenicity and carcinogenicity are the main effects on mammalians (Kima et al., 2013). When present in wastewater, high molecular weight PAHs molecules are slowly biodegraded in both activated sludge and compost piles and this limits their disposal for landfilling and for agricultural purposes. On the other hand low molecular weight PAHs (≤ 3 aromatic rings) are not efficiently retained in the sewage sludge during the treatment steps (Charalabaki et al., 2005), therefore they can be found in the effluents of the WWTPs and this increases their diffusion in the environment (Charalabaki; 2005). Due to their high hydrophobicity, PAHs can be absorbed and can accumulate into particulates and soil fractions from contaminated water stream. The sorbent organic matter, which gathers these compounds can eventually sink and be a source of further pollution. This phenomenon has potentially harmful consequences for all the biota living downstream from the sinking. Furthermore, when retained in the soil or in aquatic organisms, these compounds can undergo a process of biomagnification, enter the food chain and increase the risk to human health (Ramesh et al., 2004; D'Adamo et al., 1997). Phenanthrene is one of the most common PAHs with low molecular weight detected in air, soil, wastewater streams and treated wastewater effluents (Wloka et al., 2013, Chang, 2006, Charalabaki et al., 2005). Processes leading to the formation of PAHs are common in urban and industrial areas. Therefore the occurrence of these pollutants is widespread in soil, surface water, air and wastewater. Due to the level of information and specific high occurrence in wastewater, phenanthrene has been chosen as reference compound for this study. Although it has not been proved to be carcinogenic for human beings, phenanthrene is often considered as analytical marker for evaluate the exposure to PAH sources, in relation to other more dangerous PAHs (Kuusimaki et al.,2004, Srogi 2007).

1 Introduction

In Table 10 are listed examples of PAH occurrence in soil. It is possible to observe that the highest concentration of PAH in soil is found within industrial areas and more specifically where coal burning processes are performed. (Placha et al., 2009, Hussar et al., 2012). Furthermore, among all the PAHs, the reviewed studies confirm that phenanthrene is found to be the most common pollutant in both soil, water and sewage sludge (Table 10, 11, 12).

Table 10 Occurrence of PAHs in soil reported by literature references (Placha et al., 2009, Placha et al., 2010, Okedeyi et al., 2012, Hussar et al., 2012). dw= dry weight

Soil	∑PAHs (mg/kg dw)	Phenanthrene (µg/kg dw)	Reference
Soil near the Tiefa Coal mine, China	0.06 – 5.64	-	Liu et al.,2012
Plant for production of cement, Italy	0.1 - 96	5.3 - 33	Orecchio 2009
Coal tar refinery, Czech Republic	0.8 - 10	-	Placha et al.,2009
Forest soil near coal tar refinery, Czech Republic	0.7 - 79	-	Placha et al.,2009
Seine river basin, urban area, France	0.005 – 0.3	132	Motelay-Massei 2003
Seine river basin, industrial area, France	0.009 – 0.6	254	Motelay-Massei 2004
Urban area of Minsk, Belarus	0.6	-	Kukharchyk et al.,2013
Coal fired power plants, South Africa	9.7 – 6.1	-	Okedeyi et al.,2013
Industrial area of Chattanooga, Tennessee, USA	0.6 – 20.8	50 - 6600	Hussar et al.,2012

In water the occurrence of PAHs depends on the level of industrialization and urbanization of the monitored areas. The road run off, the industrial and household activities are mainly involved in the surface water pollution. A higher density of population and a high level of industrialization causes higher concentration of PAHs in the water bodies compared with areas where the industrialization is lower and better controlled (as for soil in Table 10).

Table 11 Occurrence of PAHs in surface water. Comparison between high population density areas and low density areas.

Surface water in urban and industrial areas	∑PAHs (µg/L)	Phenanthrene (ng/L)	Reference
Raba river, Urban and industrial area, Hungary	0.041 – 0.4	-	Nagy et al., 2013
mean from all the rivers, high density, China	0.003 – 38.1	-	Guo et al., 2012
Lake Maggiore, Italy	0.003	0.8 - 1.9	Olivella et al., 2006
Venice Lagoon, Italy	0.003	0.8 - 2.24	Manodori et al., 2006
Tväran river and Nemunas river, Lituania	0.0 8	7.7	Bergqvist et al., 2007

Also the concentration of PAHs in sewage sludge is strictly linked to the typology of the reference area. The range of concentrations is uniform between urban areas but differs largely from the small rural area described in Mansuy-Huault et al., 2009 (Table 12).

Table 12 Occurrence of PAHs in sewage sludge reported by literature references. (dw=dry weight)

Sewage sludge	∑PAHs (mg/kg dw)	Phenanthrene (µg/kg dw)	Reference
Rural district, Lorrain, France	0.8 - 60	0.08 - 6	Mansuy-Huault et al., 2009
Urban area, Poland, before composting	2 - 10	-	Oleszczuk 2008
Urban area Poland, after composting	1 - 7	800 - 5600	Oleszczuk 2009
Urban area Paris, France	14- 31	1500 - 3000	Blanchard et al., 2004
Urban area, Bejiing, China	2.4 - 26	48 - 466	Dai et al., 2007

1.1.2.3 Properties of PAHs and phenanthrene

The polycyclic aromatic hydrocarbons, also called polycyclic arenes, are organic compounds formed by the fusion of 2 or more benzene rings. There are several organic compounds classed as polycyclic aromatic hydrocarbons, but only about 16 are listed as reference compounds for the US EPA and the European Community (Table 13).

1 Introduction

PAH have a common molecular electronic arrangement proper of conjugated systems. Conjugated molecules have π electrons that are not localized in individual double or triple bonds.

All physical-chemical properties of PAHs are related to the arrangement of π electrons. The main properties are (Lee et al., 1981):

- Low Solubility
- High vapour pressure
- High octanol-water partition coefficient
- High octanol-air partition coefficients

Briefly these properties are listed in Table 13, where the structure and the main properties of these compounds are presented (Sverdrup et al., 2002; Chiou, 1985.).

Table 13 The main chemical properties of all the PAHs listed by the US EPA and European community as priority substances (Sverdrup et al., 2002). Molecular weight (MW), water solubility (WS), octanol-water partition coefficients (L/Kg), vapour pressure (Pa) and structural formulas. Phenanthrene is highlighted as reference compound for this study.

Name	MW	WS (mg/L)	Log K_{ow}	Log K_{oa}	V_p / Pa	Structural formula
Benz(a)anthracene (BaA) $C_{18}H_{12}$	228.3	0.0094	5.8	10.28	2.80×10^{-5}	
Benz(a)pyrene (BaP) $C_{20}H_{12}$	252.3	0.0016	6.20	11.35	7.32×10^{-3}	
Benz(b)fluoranthene (BbF) $C_{20}H_{12}$	253.3	0.0015	6.4	11.34	6.67×10^{-5}	
Benz(k)fluoranthene (BkF) $C_{20}H_{12}$	252.3	0.0008	6.40	11.37	1.29×10^{-7}	
Crysene (Chr) $C_{18}H_{12}$	228.3	0.0020	5.80	10.30	8.31×10^{-7}	

1 Introduction

Name	MW	WS (mg/L)	Log K_{ow}	Log K_{oa}	V_p / Pa	Structural formula
Pyrene (Pyr) C16H10	202.2	0.13	5.2	8.86	6.00×10^{-4}	
Dibenz(a,h)anthracene (DBahA) $C_{22}H_{14}$	278.4	0.0025	6.50	12.59	1.27×10^{-7}	
Indeno(1,2,3-c,d)pyrene (IPyr) $C_{22}H_{12}$	276.3	0.0002	6.70	12.43	1.67×10^{-8}	
Naphthalene (Nap) $C_{10}H_8$	128.2	31.0	3.32	5.19	1.13×10^{1}	
Acenaphthene (Ace) C12H10	154.2	3.9	3.94	6.44	2.87×10^{-1}	
Acenaphthylene (Acy) C12H8	152.2	16.1	4.07	6.46	8.91×10^{-1}	
Anthracene (Ant) C14H10	178.2	0.0434	4.50	7.70	8.7×10^{-4}	
Benzo (g,h,i)perylene (BghiP) C22H12	276.33	0.0003	6.6	12.55	$1,33 \times 10^{-8}$	
Fluoranthene (Fla) C16H10	202.2	0.26	5.2	8.81	1.23×10^{-3}	
Fluorene (Flr) C13H10	166.2	1.69	4.2	6.85	8.0×10^{-2}	
Phenanthrene (Phe) C14H10	178.2	1.3	4.6	7.64	1.61×10^{-2}	

1 Introduction

PAHs can be divided according to the molecular mass, in low molecular weight (LMW) having a number of benzene rings up to three and high molecular weight (HMW) having more than three benzene rings.

The LMW PAHs show higher solubility and volatility than the HMW PAHs. In addition the spatial arrangement of the aromatic rings within the molecule, provides highly different physical-chemical properties. PAHs can be formed by linearly fused rings or can present an angular disposition of the aromatic rings.

Greater thermodynamic stability arises from the delocalization of the π- electron density, so that liner fused PAHs are thermodynamically less stable than angular PAHs. Furthermore the higher are molecular weight and molecular arrangement, the stronger are the hydrophobicity and the electrochemical stability of the molecule and according to this principle, also the solubility, the vapour pressure, the chemical reactivity and the photochemical reactivity are significantly lowered.

As previously mentioned, higher molecular weight PAHs show stronger hydrophobicity and in the environment, this is in correlation with a lower biodegradability of the molecule. In fact the high molecular weight PAHs have a higher octanol-water partition coefficient (>4.5) and show a lower solubility in water , so that they are sequestered on particles of solid organic matter (SOM) and are less available for microbial communities to be degraded (Singh, "*Microbial degradation of xenobiotic*",Springer, 2012).

The vapour pressure decreases with increasing molecular weight and this causes the increase of the boiling temperature. Naphthalene (low molecular weight) has a vapour pressure equal to 11 Pa and a boiling point equal to 218 C, while BaP (high molecular weight) has a vapour pressure much lower around 7×10^{-2} Pa and a boiling point equal to 495 C.

PAHs are chemically classified as rather inert compounds, however when they do react, tend to retain their conjugated ring system and this arises the formation of derivate compounds by electrophilic substitution rather than addition reactions.

1 Introduction

PAH compounds can absorb light both in the UV and in the visible regions of the spectrum. The absorption of light energy, leads to the excitation of the molecule. Firstly, the electrons of the molecule achieve upper energy states, then they return to the ground energy state gradually losing their energy. During this step, the PAH molecule can release the energy as photon (fluorescence or phosphorescence). However, there are other possible pathways involved in this loss of energy from the excited state. Two of them are the energy transfer and the electron transfer to other molecules, which are present in the surrounding.

As previously mentioned, combustion processes (e.g. waste incineration) of organic matter is the main category of processes leading the distribution of PAHs in the overall environment. The industrial liquid effluents and the combustion gases are the main carriers which spread the PAH into the atmosphere, soil and aquatic systems. PAHs in the gas phase can undergo different processes depending on their volatility. Gas-particles partitioning is the main process that drives the distribution of PAHs in the atmosphere. Semi-volatile PAHs and all higher molecular weight PAHs are likely to be absorbed to particles of organic matter and adsorbed to soot, while volatile PAHs exist mainly in the gas phase (Keyte et al., 2013). Through the deposition of atmospheric particles, PAH can be accumulated in soil and plants. When PAHs occur in aqueous environments upon transport through industrial effluent, their solubility determines their distribution between the organic particles, the bottom sediments and the water column. The diffusion of PAH from the water streams into the soil matrix is another process increasing their accumulation in the environment. Higher molecular weight PAH are most likely found absorbed into the bottom sediments of water streams, while lower molecular weight PAHs are mostly absorbed on particles in the water phase (Readman et al., 1984). The fate of PAH compounds in the environment can be divided in four main categories of processes: abiotic oxidation, photo oxidation, sorption into organic matter and biodegradation. Temperature, turbidity, concentration of dissolved and particulate material and the nature and concentrations of microbial populations are the properties of the eco-system, which influence these transformation processes. PAHs can react with O_3, NO_3 radicals, NO_2, OH radicals, peroxides, sulphur oxides, chlorine and these reactions produce many different derivative compounds. All this processes rise much interest because the derivative compounds produced (e.g. oxygenated PAHs or polychlorinated-PAHs) display a significant toxic potential, in some case higher than the parent PAHs (Kochany and Maguire 1994). Linear PAHs undergo Diels-Alder reactions at the 9,10 positions and in presence of light might form endoperoxides when oxygen is available and photodimers in absence of oxygen. Angular PAHs are not subjected to Diels-Alder reactions and photodimerization, but can form endoperoxides (e.g 9,10-quinones) in presence of oxygen and under irradiation of light (Lee et al.,1981). These reactions increase the toxicity of PAHs, as proved by several studies on different organisms such as benthic invertebrates, aquatic vertebrates, plants and also mammals (Arfsten et al., 1996). The presence of the skin in mammals allows the filtration of most of the light, limiting the toxicity to the out layer surface, causing photoallergies or non-immunologic induced skin reactions. Acute phototoxic effects have been observed after co-exposure of mice to PAHs and direct UV light. Mice painted with benzo(a)pyrene and exposed to sunlight for 30 minutes and 1 hour developed erythema and acute dermatitis (Arfsten et al., 1996). Furthermore, since the photo oxidized species are more water soluble than the parent PAHs, organisms could be exposed to higher concentrations of photomodified PAHs than the parent PAHs. This presents a greater toxic risk, because, as previously mentioned, oxidized PAHs are known to be more reactive and biologically damaging than the parent compounds (McConkey et al., 1997).

The conjugated nature of these molecules influences also their spectroscopic properties. Bigger conjugated systems absorb UV/Vis light at longer wavelength. This is easily observed for linear PAHs, but is valid for all the PAHs (figure 8, Reusch, 2013).

1 Introduction

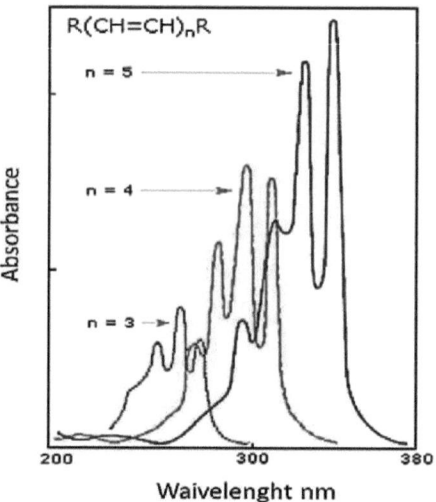

Figure 3 Absorbance spectra of the three   **Wavelength (nm)**   : hydrocarbons in deionized water with a concentration of 0.5 mg/L (Reusch 2013). n indicates the number of aromatic rings composing the PAH compound characterized by each spectrum.

The absorption spectra of all PAHs are characterized by three main absorption bands: α, β, p. In a fluorescence emission spectrum each band is correlated to a different electron energy transition to the exited state after a light excitation event. These three bands, which are common for all PAHs, shift to the longer absorption wavelength with increasing molecular weight of the PAH compounds. Therefore, it is possible to identify different PAHs by using the UV-absorption spectra. Furthermore the lower molecular weight PAHs need shorter excitation wavelength to be detected by UV light and higher molecular weight PAHs can be fluorescent at longer excitation wavelength. This phenomenon has been called annellation principle by E. Clar, who studied the correlation between the π electron density and the spectroscopic properties of aromatic compounds (Clar, 1964).

As previously mentioned PAH compounds represent a danger for human health. The carcinogenicity of PAHs has been largely investigated and has been found that the most potent carcinogens among PAHs are the ones resulted from reactions of addition or redox reaction on specific areas of the molecules. In fact, some PAH can have specific regions where they are more active. These regions are called "bay region" and "k-region". Metabolites like arene oxides, hydroxyl, dihydroxy drivates, dihydrodiols and quinines are among the most carcinogenic pollutants (Jerina et al., 1976).

At 21°C temperature, phenanthrene has a water solubility between 1.2 and 1.5 mg/L (Eastcott et al., 1988) an octanol/water partition coefficient of around 4.3 - 4.6 (Chiou et al.,1998) and a diffusion co-efficient in water of 7×10^{-6} cm²/s (Schüth, 1994). It is a semi-volatile organic compound with a vapour pressure of 6.8 $\times 10^{-4}$ mmHg. Phenanthrene shows UV-absorbance below 300 nm and fluorescence, between 345 nm and 382 nm with an excitation wavelength of 290 nm (PubChem-NCBI).

Although PAHs are commonly recognized as potential carcinogens, there are not experimental proves that phenanthrene is carcinogenic for mammalians, and more specifically for human beings (US EPA 2001). However Simmon et al., 1979 reported an oral median lethal dose (LD_{50}) of 750 mg/kg for mice. Single doses of 100 mg/kg/day of phenanthrene administered by gavage for 4 days suppressed carboxylestrase activity in the intestinal mucosa of rats, but did not produce other signs of gastrointestinal toxicity. Phenanthrene had no effect on hepatic or extrahepatic carboxylesterase activities (Nousiainen et al., 1984).

The importance of phenanthrene within environmental issues is represented by its high co-occurrence in water, soil and sewage sludge with more dangerous PAH compounds. Therefore phenanthrene is listed as priority substance for its referential role in the environmental monitoring of polycyclic aromatic hydrocarbons.

Phenanthrene is also largely used for specific industrial products. It is involved in the synthesis of explosives, dyes, drugs and steroids (fact sheet by US Environmental Protection Agency, web, 2014). Phenanthrene based alkaloids are produced as antitumor substances (Wei et al., 2006), phenanthrene based tylophorine derivatives are used as antiviral agents, e.g. against the Tobacco mosaic virus (TMV) in plants (Wang et al., 2010). In the production of photovoltaics, specifically for bulk hetero conjunction (BHJ) solar cells, phenanthrene is a main component of the conjugated polymers. In the sintering industrial applications for the processing of thermoplastics, phenanthrene is a component of the sinterable polymers (e.g. US patent 2400099 A).

1.1.2.4 Octanol-water partition coefficient (Kow)

In the environment all chemicals are continuously transported and redistributed between solid, liquid and gaseous phases. The rate of transportation from one phase to another depend on the affinity of each chemical for each phase. In fact the actual issue about pollutants is their affinity for different environmental compartments. Risk assessment procedures are based on the transport of pollutants at water-solids, water-gas and gas-solid interphases. In this regard for each pollutant it is important to define specific parameters, which could describe their partitioning between these phases. These parameters are called partition coefficients. The general partition coefficients (K_p) are calculated by the ratio between the molar concentrations of the chemical in two immiscible phases (phase 1 and phase 2) at the equilibrium as follows:

$$K_p = \frac{c_1}{c_2} \qquad (Eq.1)$$

Solid phases refer to inorganic and organic matter. Part of the organic matter are soil, living organisms and other matrices such sediments, sludge and microbial biofilms. The aqueous phase and the gaseous phase are the most important phases for the transport of pollutants through the environment, while the solid phases play often the role of sinks and sources. A specific pollutant can be absorbed and accumulated into a solid phase and then, under different environmental conditions, can be desorbed back into the environment within the gas or liquid phase. In order to investigate the occurrence of a pollutant in the environment it is necessary to know its water-solvent phase partitioning. Since long time the partitioning of organic chemicals between pure water and n-octanol and the resulting octanol-water partition coefficient (K_{ow}) has been widely used as first reference parameter for predicting the partitioning of organic chemicals between soil and water and between living organisms and water (Kenaga and Goring, 1978). The higher the K_{ow}, the more hydrophobic is the compound. Therefore this partition coefficient is a measure of the water solubility of chemicals. In fact Log K_{ow} (L/kg) values are generally inversely related to aqueous solubility and directly proportional to molecular weight (USGS, website, 2014). High values of K_{ow} indicate the tendency of a pollutant to be partitioned mostly into biota and soil rather than in water phases and therefore this parameter represent an important reference (Miller and Wasik 1985). Few methods are applied for measuring this parameter for different compounds, but it is possible to calculate it using an equation proposed by Chiou and co-workers (Chiou et al., 1982) and reviewed in Miller and Wasik 1985. This equation has been tested for 34 organic pollutants, resulting a predicting error less than one order of magnitude between the calculated and the measured values. Therefore it is possible to predict the octanol-water partition coefficient of a pollutant, knowing its water solubility and the activity coefficients of each compound in the two immiscible phases.

Given the K_{ow} as follows:

$$K_{ow} = \frac{(\gamma_{wo} v_{wo})}{(\gamma_{ow} v_{ow})} \tag{Eq.2}$$

The octanol-water partition coefficient is expressed as follows:

$$Log\ K_{ow} = Log\ Ws - \log Vo - \log \gamma o + \log \frac{\gamma w}{\gamma wo} \tag{Eq.3}$$

Where γ_w is the activity coefficient of the chemical in water, γ_{wo} the activity coefficient in water saturated with octanol, γ_{ow} is the activity coefficient of the chemical in n-octanol saturated with water, v_{wo} is the molar volume of water saturated with octanol (m³/mol) and v_{ow} is the molar volume of octanol saturated with water (m³/mol) V_o is the molar volume of the organic solvent and Ws is the water solubility of the chemical.

In 1989 De Bruijn and colleagues (De Bruijin et al., 1989), provided a good review about the methods used for the measurement of the octanol-water partition coefficient. The classical method for measuring the K_{ow} is the flask-shaking method. The chemical is mixed with an appropriate 1-octanol/water mixture and shaken for some given period during which equilibrium between both phases must be achieved. After both phases are allowed to separate, the concentration of the chemical in both phases is determined. Although this method has been widely used in the history, it has been proved to be not accurate for the determination of the partitioning of the more hydrophobic compounds due to the formation of water/octanol emulsions which interfere with the partitioning procedure. Therefore other methods have been developed. Brooke et al., 1986 developed the so called "slow stirring" method. In this case the water and the octanol phase are equilibrated under conditions of slow stirring and the formation of emulsions can be prevented, resulting in a more precise measurement of the partitioning. The octanol-water partition coefficient is one of the most important parameters for predicting the environmental mobility of pollutants among the different compartments of the environment and it is taken in this study as a reference for modeling the sorption of phenanthrene into sewer biofilms.

1.1.2.5 Analytical methods to detect phenantherene

The procedures involved in the analysis of PAH compounds are different and depend on the nature of the sample analyzed. The analysis of organic pollutants from environmental samples is based on few necessary steps: extraction, separation, concentration and detection. Water, soil and sludge are three different matrices, which require different approaches. As reported by Manoli and Samara, 1999 the most widely used extraction techniques for water samples are the liquid-liquid extractions (LLE) and the solid phase extractions (SPE). According to the protocol EDIN38407 F18 the most suitable solvent for this extractions is n-hexane and alternatives are benzene, toluene, dichloromethane and cyclohexane. For marine water samples it is recommended to use a mixture of light petroleum and diethyl ether in order to perform the extraction of PAHs (Bruzzoniti et al., 2000).

Beside the good results and the low costs, the LLE implies the disposal of large volumes of toxic solvents, of environmental concern, and relatively long time procedures. On the other hand SPE does not requires large volumes of solvents and automatic procedures can be easily settled to decrease the analysis time. SPE implies the use of special cartridges packed with silica as solid phase on which the PAH molecules are adsorbed. The recovery of the targeted compounds is carried out by solvent elution. When following a prior steam distillation step, the SPE approach represents a valid choice. However the SPE procedure is suitable for „clean water" samples, because the presence of particle-bound PAHs affects its efficiency significantly.

A more effective approach for recovery of PAH from environmental samples is the solid phase microextraction (SPME). This technique involves the use of special fibers as solid phase, coated with polydimethylsiloxan (PDMS) and immersed into the sample while mechanical agitation, stirring or ultrasonication is applied. The affinity between the fiber and the target substance will allow the efficient recovery even though on complex environmental samples. Since latest '90s the column extraction of PAHs from water samples bythe use of immunosorbents has gained interest. The antibody is immobilized on a silica support and used as affinity ligand in order to bind the target analyte from the aqueous phase. Bouzige et al., 1998 proposed a more effective immunosorbent extraction method using an anti-fluorine assay followed by liquid chromatography with diode array detection (LC-DAD) on sediment and sludge samples. Although good results are summarized in literature, the efficiency of this technique is affected by the complexity of the sample, hence solvent extractions are preferable.

In this regard a broad range of methods are available since different types of samples require different solvents and conditions. For instance in case of marine water samples the best results are achieved by light petroleum diethyl ether mixture, followed by determination procedures based on high-performance liquid chromatography (HPLC). In case of soil, sludge or slurry samples the approaches are different.

The most widely used techniques are (Northcott and Jones, 2000; Song et al., 2002; Semple et al., 2003):

- Batch solvent shaking extraction.
- Soxhlet extraction.
- extraction after ultrasonic treatment.
- SPME followed by gas chromatography/mass spectrometry (GC/MS)

Alternatives methods are (Camel, 2000):
- Microwave assisted solvent extraction.
- Supercritical fluid extraction.
- Accelerated solvent extraction.

After the hydrophobic organic compounds (HOC) have been recovered from the samples, the separation and the detection of the PAHs must be performed.

There are mainly two separation techniques: liquid chromatography (LC) or gas chromatography (GC).

The LC consists of a polar stationary phase and the substances are separated by the different affinity for the polar support. In the case of PAHs is common the use of a reverse LC method, in which the stationary phase is a nonpolar matrix and implies the use of an elution solvent to recover the molecules adsorbed on the stationary phase.

Nowadays the most common LC technique is the high pressure liquid chromatography (HPLC), that allows higher separation efficiency and higher resolution.

The GC for PAHs is usually employed with fused silica capillary columns coated by a liquid phase (e.g. methyl silicon).

These two kind of separation are coupled with different detection techniques:
- Flame ionization detection (FID).
- Fluorimetric detection (FLD).
- Ultraviolet detection (UVD)
- Mass spectrometry detection (MS).

The UVD can be performed by UV absorption or by photodiode array (PDA). The UV absorption is most widely used due to higher sensitivity (Robards et al., 1994).

UVD and FLD are most widely used coupled with LC, while MS might be used together with both LC and GC. According to the EDIN38407F18 and the US EPA method 610 the most suitable combinations for the wastewater analysis are:

- HPLC-FLD
- HPLC -UVD or-FLD
- GC-FID

According to current literature, the detection of phenanthrene by fluorescence spectroscopy on soil or sludge samples has not been applied, while for aqueous environmental samples has been performed only using pulsed laser technologies and by analyzing the fluorescence lifetime of the target molecule (Meidinger et al., 1993).

Furthermore fluorescence detection with pulsed laser followed by HPLC has been reported to be a valid alternative to the classical approaches (Manoli, 1999).

Although several clean up steps are needed to analyze the fluorescence of PAHs from environmental samples, Campiglia et al. in 1995 proposed the use of a laser excited synchronous luminescence device in order to trace Benzo(a)pyrene in various environmental samples. This technique has been proposed for its potential application for remote in situ sensing.

Other useful approaches can be (Patra, 2003):

- Excitation emission matrix fluorescence (EEMF)
- Synchronous fluorescence scan (SFS)
- Selective fluorescence quenching (SFQ)
- Time-resolved fluorescence spectroscopy (TRF)
- Phase-resolved fluorescence spectroscopy (PRFS)
- Fluorescence correlation spectroscopy (FCS)

Amongst all the methods mentioned, the SPME extraction or the liquid-liquid extraction coupled with the GC-MS analysis is the most suitable approach for the detection and quantification of environmental samples, although it can't be performed for in situ measurements. On the other hand the fluorescence spectroscopy methods, when applied on suitable surfaces, have the advantage to be performed on field using fiber optic devices for measurements. This methods would allow faster analyses than the other two procedures previously described and would increase the benefit of an environmental monitoring strategy. Furthermore since hydrophobic pollutants, such as phenanthrene, display low water solubility, the analysis of water samples is not a reliable approach for their detection in the environment. Being mostly bound to sediments, biofilms and particles moved by the current, hydrophobic compounds are not easely detected when punctual sampling of the flowing water phase is carried out. In this regard it is worth to exploit those organic matrices, which can easily absorb this compounds from the water phase during their transportation, since the permanence of the pollutant is significantly enlarged within these matrices rather than in the aqueous phase. In this regard microbial biofilms might be suitable for detecting hydrophobic pollutants, such as PAHs in the environment. Microbial biofilms grow onto surfaces directly in contact with the contaminated aqueous systems and are proven to absorb hydrophobic pollutants. Although the use of microbial biofilms as bio-monitors for pollutants has been already reviewed, there is a substancial lack of knowledge about the real extent of the memory effect.

1.2 Microbial biofilms

Biofilms are the main microbial form of life on the planet Earth (Flemming and Wingender, 2010). A biofilm is composed of microbial cells immersed in a matrix of self-produced polymeric substances such as proteins, lipids, polysaccharides, and nucleic acids, (Flemming and Wingender, 2010; Dunne, 2002; Lazarova and Manem, 1995; Frølund et al., 1995). Biofilms are ubiquitous and grow at all sort of interfaces (liquid/air, solid/air, solid/liquid) in lakes and streams. Thickness, water content, total density, polarity, chemical and microbial composition are among the most important parameters in the study of biofilm formation. Over recent decades, the scientific community substantially increased its interest in microbial biofilms for their role as active mediators of several environmental processes. Biofilms growing at the water/liquid interface can sequester particulate and dissolved matter present in the water, so that the acquisition of nutrients by the biofilm microorganisms is facilitated

1 Introduction

(Strathmann et al., 2007). In addition, through the sorption of solutes from the contact phases, biofilms play an important role on the fate and distribution of the pollutants in the environment (Flemming, 1995). The term sorption refers both to adsorption, absorption and desorption. Adsorption implies the retention of a solute on the surface of the particles of a material. Absorption in contrast involves the retention of a solute within the interstitial molecular pores of such particles. (Strathmann et al., 2007). There are 3 main sorption sites in a general microbial biofilm matrix:

1. EPS (Extracellular polymeric substances) consisting of polysaccharides, proteins, nucleic acids and lipids providing hydrophilic and hydrophobic regions within the matrix.
2. Cell walls and lipid membranes again providing charged and not charged areas.
3. Cytoplasm as separate water phase

Each one of these sites displays specific physico-chemical properties and affinity for different kind of molecules; for instance, it has been already reported that heavy metals are strongly retained at the level of the cell walls, while bigger molecules like BTX (Benzene, toluene and xylene) are retained mainly inside the EPS layer (Spaeth et al., 1998). Furthermore, the physico-chemical properties as the microbial composition of the biofilms are closely associated with the environmental conditions of which they are exposed to. For example, under dehydrated conditions and being deprived of water, biofilms appear as crusts and show significantly different mechanical properties (Garcia-Pichel et al., 2003).

The knowledge attained so far about this kind of matrices brought significant improvements in many technical and scientific applications („Microbial Biofilms: Current Research and Applications "by Lear and Lewis, 2012).

1.3 Fate of PAHs in Sewers and role of sewer biofilms

Sewers are the main infrastructure for the transport and collection of wastewater in all European urban areas. Sewers can be represented by a simple model as composed of six different compartments (Figure 4):

- The wastewater suspension
- The particulate matter present in the aqueous phase.
- The bottom sediments
- The gas phase
- The fats, oils and grease portion (FOG), which accumulates on the surface of the aqueous suspension and at the sewer sidewalls.
- The microbial biofilms, which grow on the sewer surfaces and on the top layers of sediments, both in contact or not with the water stream.

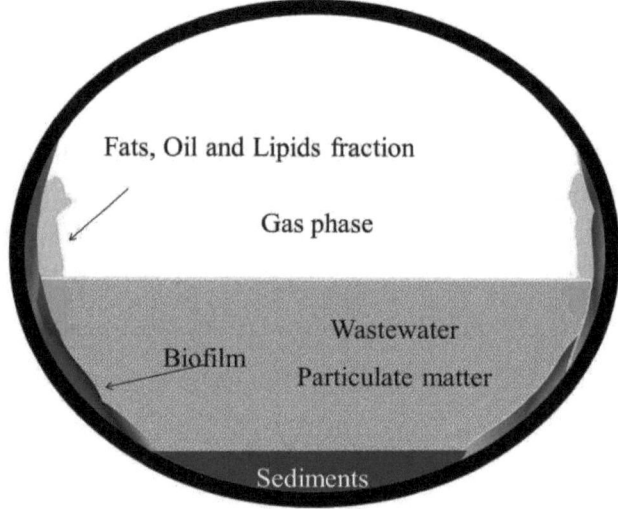

Figure 4 Section of a sewer pipe. The sketch shows the main partitioning phases, which are present in this environment.

In such an environment, hydrophobic pollutants (e.g. PAHs) encounter different fates. Low molecular weight PAHs, composed of 1-3 aromatic rings, with a Log K_{ow} between 2 and 5 are mainly found in the water solution, constantly moving from the dissolved aqueous phase to the organic particles which are suspended in the water column and vice versa (Blanchard et al., 2004). The low molecular weight PAHs can also move from the aqueous phase to the gaseous phase depending on their molecular weight. Generally PAH with molecular weight lower than 234 g/mol (Gigliotti et al., 2002) and Henry's law constant greater than 1×10^{-4} atm m^3/mol (Manoli and Samara 1999) are mainly found in the gaseous phase rather than in water. In fact the lower is the molecular weight of a PAH compound, the lower it is the water-air partition coefficient (Log K_{wa}) and therefore the more volatile is the compound. Another important process for low molecular weight PAHs is the bacterial biodegradation. In sewers, this process is significant only for compounds with less than 4 aromatic rings (Blanchard et al., 2004).

On the other hand the volatilization and bacterial biodegradation are not a significant removal processes for high molecular weight PAHs (more than 5 aromatic rings) present in wastewater. The higher Log K_{ow} and the lower solubility of these compounds allows their prevalent sorption into organic matrices such as sediments, fats, oils and lipids portions and of course microbial biofilms. Only a smaller amount of these compounds are found in the dissolved phase of the water column, accounting less than 20% of the total (Blanchard et al., 2004).

In general the sorption of hydrophobic compounds into organic matrices is described by the organic carbon partition coefficient (K_{oc}), which is calculated through the K_{sw}:

$$K_{sw} = \frac{q_s}{C_w} \qquad (Eq.4)$$

and as follows:

$$K_{oc} = \frac{K_{sw}}{f_{oc}} \qquad (Eq.5)$$

Where K_{sw} is the equilibrium distribution factor of the compound between the water phase and the sorbent material, q_s is the uptake of the compound into the sorbent phase and C_w is the concentration of the pollutant in the water phase. f_{oc} is the fraction of organic carbon content of the sorbent material (Chiou et al.,1983, Wicke et al.,2007).

In order to predict the sorption of PAHs into organic matter, such as soil, sediments, biofilms or more general organic particles, two main models have been elaborated.

Both of them relate the octanol-water partition (K_{ow}) coefficient with the organic carbon partition coefficient (K_{oc}).

For substituted PAH (e.g. polychlorinated PAHs) with a Log K_{ow} ranging from 2-5, Chiou et al., 1983 proposed the following linear relationship:

Log K_{oc} = 0.9 Log K_{ow} − 0.543 (Eq.6)

For PAHs with higher values of Log K_{ow}, Karickoff et al., 1979 proposed as follows:

Log K_{oc} = 1.0 LogK_{ow} − 0.21 (Eq.7)

These equations are important to model and predict the partitioning of PAHs in the open environment, but can be used in order to describe their partitioning also in sewers.

Also the deposits of fats, oils and lipids play a significant role in the distribution of PAHs in sewer systems. It is widely known that these substances display adhesive properties, which allows them to accumulate on the sewer walls and often pose a risk of sewer overflows due to blockage events (He et al., 2013). In numerous works the significant increase that these substances gain on the sorption and the retention of PAHs into biota, food and soil has been pointed out (Bruner et al., 1994; Zhang and Tao, 2009; Moret and Conte, 2000). Therefore in this environment, where they are an important portion of the organic matter, they must be taken into great consideration.

An important linear relationship between the lipid-water partition coefficient (Log K_{lw}) and the octanol-water partition coefficients (K_{ow}) of PAHs has been proposed by Zhang and Tao, 2009:

Log K_{lw} = 1.23 Log K_{ow} − 0.78 (Eq.8)

Also in this case of study the octanol-water partition coefficient linearly influence the sorption of different PAHs on lipid rich matrices.

1.3.1 Sewer biofilms

The abundance of organic and inorganic compounds present in waste water allows the formation of biofilm matrices in sewers (Jahn and Nielsen, 1998). The sewer biofilm system differs from the other biofilm systems in two important ways (Nielsen et al., 1992):

- Very high organic loading
- High shear stress at the biofilm surface

These two factors cause the formation of an extremely rough and thick biofilm matrix with high content of organic substances. Therefore the transportation of substrates and the EPS composition are significantly affected compared with different biofilm models. The total content of EPS produced by bacteria in sewer biofilms can differ highly depending on different environmental parameters. Sheng et al., 2010 summarized them as substrates, nutrient content, metal concentration, shear rate, aerobic and anaerobic conditions.

Although these factors are really difficult to be combined in order to create a suitable conceptual model, some investigation has been carried out and useful reference considerations have been suggested. Jahn et al., 1998 analyzed the biofilms from three different sewer lines from the Aalborg area in Denmark. The cell biomass was found to be the minor fraction of the total composition equal around to 2-12% of each sample, whereas proteins (50%), humic substances (1-10%), polysaccharides (30%) and uronic acids (5%) were found to be the most relevant components. Lipids and nucleic acids are not considered as main component of the EPS in this study. Adav and Lee (Adav and Lee, 2008) studied the EPS composition of waste water sludge samples from a local municipal WWTP in Taipei, Taiwan and found that the content of lipids was equal to 8%-10% and the DNA was equal to 0.2% of the total volatile suspended solids. The polysaccharide/protein ratio (PN/PS) is also a reference parameter for investigations. Jahn and Nielsen 1998 calculated the PN/PS ratio equal to 0.6, confirmed by Zhang and Tao, 1999, which estimates it between 0.5 and 0.7 for biofilms grown in rotating angular reactors. Adav and Lee, found the PN/PS ratio equal to 0.9. So it seems likely that in sewer biofilms the proteins content is higher than the polysaccharide content. This specific characteristic needs to be further considered in regard to the sorption of organic pollutants by sewer biofilms.

1 Introduction

According to Jorand et al., 1998 and Celmer et al., 2008, the hydrophobic portion of a microbial biofilm is represented by proteins, therefore sewer biofilms are likely expected to have strong hydrophobic properties due the higher content in protein and humic substances than other biofilm systems.

Using molecular techniques for characterization of the microbial communities inside sewers, Vinke et al., 2001 found that bacterial and fungal communities are predominant, though bacteria are the main colonizers especially below the water level.

In this regard, the main phyla of bacteria in sewer biofilms are alpha, beta and gamma proteobacteria, acidobacteria and actinobacteria. (Satoh et al., 2009; Vincke et al., 2001).

1.3.2 Sorption properties of Biofilms

In aquatic environments, the fate of pollutants is strongly dependent on the sequestration process occurring at the solid-liquid interface. Biofilms cover most of the solid surfaces and therefore their sorption properties have an important role for the fate of organic and inorganic pollutants in sewers. In figure 6 is presented a general sewer model with all its sorptive compartments.

As previously mentioned (page 36, Figure 4), in sewers microbial biofilms represent one of the main sorbent phases for organic and inorganic substances. In biofilms the cells and the EPS layer represent important sorption sites (Flemming and Wingender, 2010).

1 Introduction

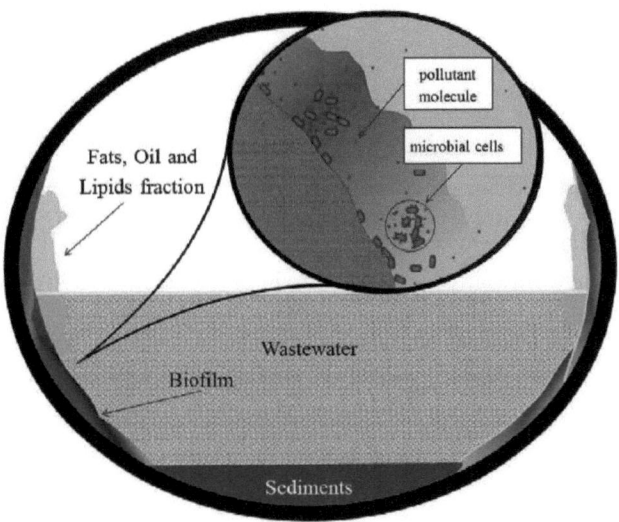

Figure 5 Sketch of a sewer and a sewer biofilm layer. The biofilm grows on solid surfaces (substrata) and is composed by EPS and microbial cells.

In this regard, two main observations must be evaluated:

1. Microbial biofilms are complex, dynamic and structured matrices. Each component of the matrix displays different sorption preferences, capacity and properties (Flemming 1995).
2. Microbial biofilms act both as sink and source of pollutants in aqueous environments.

The biofilm matrix contains several sorption sites (Table 14), such as extracellular polymeric substances (EPS), cell walls, cytoplasmatic membrane and cytoplasm. The main components of the dry mass of a biofilm are the EPS and the microbial cells. The EPS mainly comprise polysaccharides, protein, humic substances and nucleic acids. However the main component of a biofilm is water, which can reach over 98% of the total wet weight, while EPS represent the 1-2% w/w (Flemming 1995, Spaeth et al., 1998). The sorption and accumulation of heavy metals in biofilms has been largely reported and the cellular fraction, more specifically the cell wall have been identified as the main sorption site for this class of pollutants (Spaeth et al., 1998, Gutekunst 1989). The main processes governing the sorption of metal ions onto cell walls are ion exchange reactions, precipitation and complexation. In activated sludge and hence in sewer biofilms, the main process seems to be the ion exchange reaction (Sheng et al., 2010). Although the cell fraction seems to provide stronger affinity for metal sorption in surface water biofilms, Liu and Fang 2002 reported that in the wastewater framework, the hydrogen producing sludge and the sulfate-reducing bacteria (SRB) biomasses show EPS content of electrostatic binding sites 20-30 fold those reported for the bacterial cell fraction. In this case the EPS might play a major role in the sorption of heavy metals from the waste water solution.

Table 14 Sorption sites in biofilms (Flemming and Leis 2002)

Sorption sites in Biofilms
EPS (including capsules) mainly consisting of polysaccharides and proteins.
Charged groups, for example $-cCOO^-$, $-SH^-$, $-SO_4^{2-}$, $-H_2PO_4^{2-}$, $-NH_4^+$, $-NRH_2^+$ etc..
Apolar groups for example aromatics or aliphatics such as found in proteins; also hydrophobic regions in polysaccharides.
Cell walls:
Outer membrane of gram negative cells (lipids)
Murein or teichoic acid layer of gram negative and gram positive bacteria respectively
Cytoplasmatic membrane (lipids)
Cytoplasm

In nature microbial biofilms are occurring mostly as multispecies matrices, involving different species of microorganisms and often also different trophic groups like protozoa and metazoa (Wimpenny et al., 2000). Many strains form more than one strain-specific type of EPS, and the composition may change during their life cycle, so that the heterogeneity influences the sorption properties of biofilms (Flemming, 1995). Dynes et al., 2006 studied the sorption of chlorexidine in river biofilms composed by algae and bacteria. Algae were found to show stronger sorption capacity for the chlorexidine due to the production of extracellular lipid droplets, which increased the solubility of the target compound. The differences in the bioaccumulation of chlorhexidine by diatoms and bacteria indicate that not all species in a natural biofilm will contribute to bioaccumulation to the same degree. Consequently, the degree of bioaccumulation in a biofilm will depend on the presence and abundance of certain species (Sutherland, 2001). Let's consider polysaccharides. The composition of polysaccharides of a biofilms depends on the microbial communities, which compose that biofilm. Different polysaccharides found in microbial biofilms can display different physical-chemical properties. Algal alginates differ from bacterial alginates for the level of acetylation. Bacterial polysaccharides are more acetylated than algal alginates, therefore within a biofilm matrix, the higher amount of bacterial alginates might lead the formation of nonpolar regions, where hydrophobic compounds would likely be retained (Sutherland 2001). On the other hand, a higher amount of algal alginates would increase the interaction with metal ions present in the liquid phase. The tertiary structure of the polysaccharides is another factor, which influences the sorption properties of a biofilm. Most of the polysaccharides found in biofilms bind high amounts of water, but some of them (e.g. bacterial cellulose, mutan or curdlan) produced by different bacterial strains (e.g. *Enterobacter agglomerans*) are rich in sequences of 1,3- or 1,4-b-linked hexose and form triple helices strongly bound together, able to exclude water and create portions completely insoluble (Sutherland 2001). These sub-regions might be a suitable accumulation sites for hydrophobic compounds. Furthermore, gram positive bacteria have been proved to display greater sorption capacity for metals due to their thicker layer of peptidoglycan (van Hullebusch et al., 2003). These bacterial cells exhibit additional acidic functional groups, such as phosphoryl from teichoic acids and carboxyl groups from teichuronic acids, which also provide stronger sorption capacity for heavy metals and metal ions (Beveridge et al., 1997, Sheng et al., 2010). Environmental conditions also influence the sorption of organic and inorganic substances from the liquid phase. Low pH values increase the sorption of metal ions (van Hullebusch et al., 2003), while high pH values increase the sorption of hydrophobic compounds (Wang et al., 2002). The higher the ionic strength, the stronger will be the sorption of hydrophobic compounds, while low ionic strength increases the sorption of metal ions (van Hullebusch et al., 2003).

Contrary to heavy metals, hydrophobic pollutants are retained in the EPS layer. Already in 1998, Spaeth and co-workers reported that BTX (benzene, toluene, and the xylene isomers), strongly hydrophobic pollutant, was absorbed and accumulated at the EPS level of a biofilm grown in a sequencing batch reactor. The same results are reported also for other compounds (Singh et al., 2006). The accumulation of dichlofop-methyl in biofilm produced by microorganisms from an activated sludge culture was reported to be caused by adsorption on EPS (Wolfaardt et al., 1995). The aromatic amino acids and the highly acetylated polysaccharides present in the EPS matrix interact with the hydrophobic compounds, causing their retention and accumulation. Within the EPS layer, van der Walls forces, electrostatic interactions, hydrogen bonds, hydrophobic interactions and London forces are responsible for the cohesion between the EPS components and the sorbed organic compounds (Flemming, 1995; Mayer et al., 1999; Kim et al., 2000). This interactions are useful to the microorganisms for the acquisition of nutrients, but can also protect the cells from the exposure to toxic substances. For instance it has been reported that *Enterobacter cloacae* strains, resistant to high concentration of lead, increase the production of EPS when are exposed to contaminated environments (Naik et al., 2012). The functional groups at the level of the polysaccharide chains play a major role in the sorption of lead. In this case the EPS matrix was characterized by high content of metal binding groups such as carboxyl, hydroxyl and amide groups along with glucuronic acid. Similar results are reported by White et al., 1998 for cadmium exposure of a SRB biofilm and by Priester et al., 2006 for chromium exposure of a *Pseudomonas putida* biofilm. In all cases it was remarked that the increase of EPS production, specifically proteins and polysaccharides, seems to be defensive mechanism to protect the biofilm cell from the toxic effect of certain organic and inorganic compounds.

As previously mentioned, biofilms are dynamic systems, so their composition can change in response to the sorption of various substances such as heavy metals, antimicrobial compounds and pollutants. Along with a change in composition also the sorption properties of the biofilms will change. Schmitt et al., 1995 studied the effect of toluene on the biofilm properties of *Pseudomonas putida* strain. At low concentrations (5 ppm) the presence of toluene correlated with the increase of EPS production and at higher concentration an increase of carboxylic group was observed on the polysaccharide chains (Schmidt et al., 1995). Also the sorption of heavy metals can influence the properties of a biofilm.

Different experimental models for the sorption of organic compounds and metal ions in biofilms and activated sludge were proposed. The models are created mostly on the base of two experimental conditions: batch experiments and continuous packed bed systems. From batch experiments, the proposed models involve both Freundlich and Langmuir isotherms and the kinetics of sorption first order or pseudo second order kinetics. On the other hand the fundamental transport equations derived to model the fixed bed systems with theoretical rigor are differential in nature and usually require complex numerical methods to solve. Such a numerical solution is not usually difficult, but often does not fit experimental results (Aksu, 2005). Once the pollutants are sorbed by the biofilms, they can be released back in the liquid phase through a number of mechanisms and this pose a risk on the environment downstream the breakdown site.

The detachment of whole portions of the EPS matrix is one of the mechanisms responsible for the remobilization of pollutants in the aqueous environments. There are four types of detachment reported in literature: abrasion, erosion, sloughing and predator grazing (Morgenroth and Wilderer, 2000). The erosion is caused by shear stress, abrasion by collision of biofilm support particles. The sloughing event can be induced by rapid changes in environmental conditions such as a sudden increase in shear force, sudden depletion in oxygen concentration or nutrients. However little is known about the causes of spontaneous sloughing events and the effect of subsequent biofilm development (Garnya et al., 2009). Among all the detachment mechanisms, the sloughing involves larger portions of biofilm, compared with the other ones. Rice et al., 2005 observed that in the filamentous biofilm of a strain of *Serratia marcescens*, the sloughing off is controlled by quorum sensing in response to nutrient conditions.

The desorption is another important process which influences the remobilization of pollutants from the biofilms into the water phase. As previously mentioned, the sorption models for pollutants in biofilms and activated sludge depend on the experimental system used, whether a batch reactor or a continuous flow fixed bed reactor. The salinity and the pH of the water phase, and the occurrence of surfactants are the main environmental factors influencing the desorption of pollutants from microbial aggregates. The partitioning coefficients of organic pollutants between water and biofilm are important parameters in order to predict the fate of pollutants in aqueous environments (Wicke et al., 2007; Headly et al., 1998). Several models have been proposed but further investigation are needed in order to better understand the desorption process of either metal ions and organic pollutants from microbial biofilms (Di Fabio et al., 2013; Du Laing et al., 2009; Aksu 2005; Tsezos and Bell, 1988; Headly et al., 1998; Fan et al., 1990).

1.4 Biodegradation of polycyclic aromatic hydrocarbons

Although PAH are major pollutants for water and air, the soil is the ultimate depository of these chemicals. Among all the pollutants, PAHs are not easily biodegraded in soil under normal conditions and their persistence increases with the molecular weight. Biodegradation is meant as the destruction of chemical compounds by the biological action of living organisms. Bigger PAH molecules are more hydrophobic and due to the sequestration into the soil particles they are less bioavailable to the microbial community for biodegradation (Haritash et al., 2009; Cerniglia 1992).

Microbial degradation is the main degradation process for PAH compounds and occurs both under anaerobic and aerobic conditions (Bumpus, 1989; Yuan et al., 2001; Seo et al., 2009). The anaerobic degradation of PAHs is characterized by slower kinetics and occurs naturally since the microbial communities in contaminated soils and sediments exist under dominant anaerobic conditions. The anaerobic biodegradation of PAHs in soil and sediments has attracted the interest of the scientific community for its potential to be managed in favor of bio removal and recovery purposes of contaminated grounds (Haritash et al., 2009). However the anaerobic degradation of PAHs has been proved to occur only on smaller aromatic compounds up to 3 aromatic rings, for bigger PAHs there is not enough data available.

The breakdown of the organic carbon takes place by the biotrasformation in less complex metabolites and through the mineralization into inorganic minerals, H_2O, CO_2 or CH_4. The PAH degradation can involve both bacteria, algae and fungi. Prokaryotic microorganism degrades PAHs by an initial dioxygenase attack to cis-dihydrodiols that are further oxidized to dihydroxy products. Eukaryotic microrganisms use monoxigenase to initially attack the PAH molecule to form arene oxides, followed by enzymatic addition of water to give trans-dihydrodiols (Cernaglia et al., 1989; Figure 6).

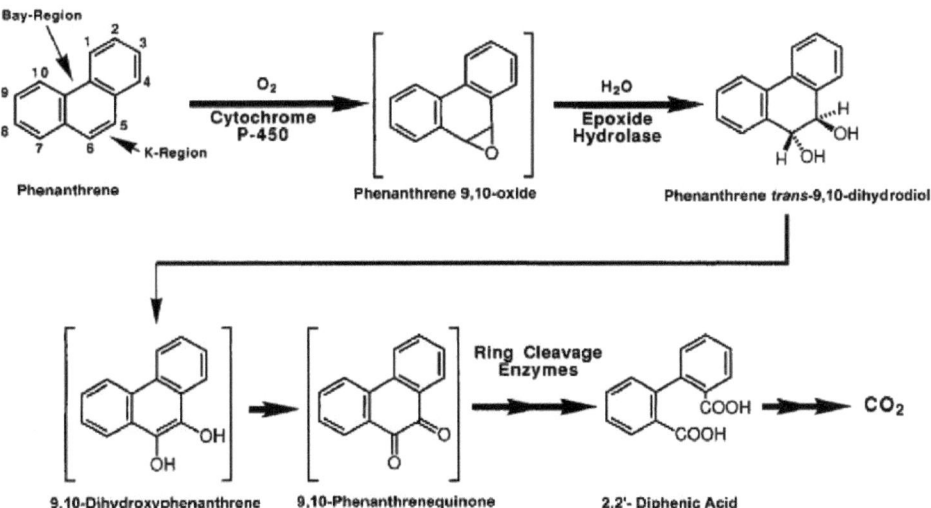

Figure 6 Proposed degradation pathway of phenanthrene by white rot fungus *P. ostreatus* (Bezalel et al., 1997)

1 Introduction

The main factors affecting the biodegradation processes are pH, temperature, availability of oxygen, microbial population, degree of acclimation, accessibility of nutrients, chemical structure of the compound, cellular transport properties and chemical partitioning in growth medium (Singh and Ward, 2004). The bacterial species able to degrade PAH compounds are summarized in table 15 and the degradation pathways are presented in Figure 7. All the strains used in these studies were sampled from contaminated sites (Haritash et al., 2009).

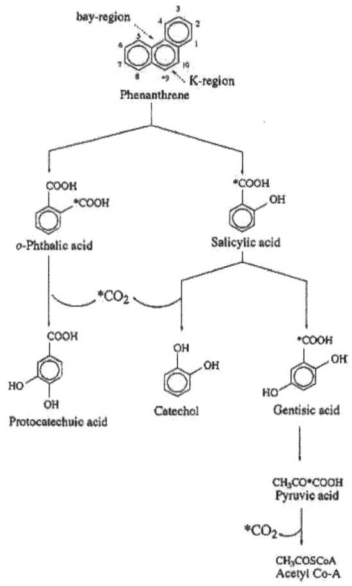

Figure 7 Proposed pathway for bacterial degradation of phenanthrene (Samanta et al., 1999)

1 Introduction

Table 15 Summary of all PAHs, bacterial species and literature references related to the biodegradation of PAH pollutants (Haritash et al., 2009).

PAH	Microrganism	Dagradation rates
Benzo(a)pyrene	Sphingomonas paucimobilis	0.96 – 1.3 µg/mL day
	Agrobacterium spp.	
	Bacillus spp.	Not available
	Burkholderia spp.	
	Pseudomonas spp.	
Pyrene	Rhodococcus sp.	0.008 mg/mL day
	Mycobacterium sp.	Not available
	Mixed culture	
	Psuedomonas/Flavobacterium sp.	Not available
Pyrene	Mycobacterium flavescens	0.56 µg/mL min
	Mycobacterium sp.strain KR2	1.25 µg/mL hour
	Pseudomonas flurescens	Not available
	Haemophilus spp.	
Phenanthrene	Paeubacillus spp.	Not available
	Rhodotorula glutinis	Not available
	Psudomonas aeruginosa	
Anthracene	Rhodococcus spp.	Not available

Among the fungal species, the ligninolytic fungi are proven to be able to degrade PAHs also under low oxygen conditions. Low molecular weight PAHs (LMW) are degraded by *Aspergillus sp.*, *Trichocladium canadense*, *Fusarium oxysporum*. High molecular weight (HMW) PAHs are degraded by *T.canadense*, *Aspergillus sp.*, *Verticillum sp*, *Achremonium sp*.

Also strains belonging to the white rot fungi (WRF) have been proved to be able to degrade benzo(a)pyrene in soil. Furthermore the presence of added surfactants to the mixture causes the increase of degradation because surfactants make PAHs easier to reach for the microorganism (Haritash et al., 2009). The most abundant fungi present in contaminated soils are yeasts. In presence of oxygen, this class of microrganisms can oxidize PAHs as alternative carbon sources. *Rhodotorula glutinis* showed high degradation rate similar to those abserved for *Pseudomonas aeruginosa*. The biodegradation of PAHs by algal biomasses is mainly achieved by employing mixed algae-bacteria microcosms. In table 10 the main studies reported by Haritash et al., 2009 are summarized.

Several studies reported the biodegadation of low molecular weight PAHs using pure microbial cultures, but no significant results proved the degradation of high molecular weight PAHs by pure cultures of microorganisms. Since different PAHs (High molecular weight and low molecular weight) are simultaneously present in contaminated samples, it seems likely that the mineralization of PAHs inside heterogeneous media (e.g. soil, wastewater or sludge), is carried out through cooperative metabolic activities of mixed microbial populations (Boonchan et al., 2000), rather than individual species of microorganisms.

The transport of PAHs inside such matrices is driven principally by the affinity that each molecule displays for the organic particles and this influences the bio-availability of the PAH molecule to the microbial populations. PAHs can be absorbed into particles, located in small pores inaccessible for bacteria, or otherwise occluded by the multitude of solid constituents. Since significant mixing forces are missing in such matrices, the diffusion of PAH molecules is reduced and their distribution is not uniform. Under these conditions, a physical barrier stands between PAH molecules and bacterial communities. The production of surfactants and EPS layers are two important strategies used by bacterial communities in order to increase the bio-availability of PAHs. Surfactants increase the diffusivity of PAHs in the water phase, enhancing their flux from the particle to the bulk liquid where the bacteria are. On the other hand, the production of EPS increases the accumulation of PAHs and provides a carbon storage for PAH-degrading communities (Johnsen and Harms, 2005).

In addition, the establishment of bacterial-eukaryotic consortia gives a further incentive to the biodegradation of PAHs in soil. In a first step ligninolitic and non ligninolitic fungi oxidize the PAHs by the action of unspecific exoenzymes and then bacteria can oxidize the more soluble by-products, through more specific PAH-dioxygenases. A first attack of the PAH molecules by fungi is more likely than a bacterial oxidation since bacterial enzymes are associated with the cell while fungal exoenzymes can diffuse better (Johnsen and Harms, 2005).

Algal populations are proved to facilitate the bio degradation of PAHs, furnishing O_2 for the metabolism of aerobic heterotrophic bacteria in wastewater maturation ponds (Haritash, 2009; Borde et al., 2003, Muñoz et al., 2003). In sewage sludge and soil mixed bacterial consortia are proved to drive the biodegradation of PAHs under both aerobic and anaerobic condition. Trzesicka-Mlynarz and Ward 1995 studied the biodegradation of BaP in soil by mixed culture of *Pseudomonas* and *Flavobacterium* species. Ambrosoli et al., 2005 showed that in soil the biodegradation of Biphenyl ($C_{12}H_{10}$), fluorene ($C_{13}H_{10}$), phenanthrene ($C_{14}H_{10}$) and pyrene ($C_{16}H_{10}$) can occur under anaerobic conditions through fermentative and respiratory metabolism of mixed bacterial communities. Chang et al., 2003 reported that in municipal sewage sludge the anaerobic biodegradation of several PAHs (phenanthrene, acenaphthene, fluorene, pyrene and anthracene) is carried out by mixed bacterial communities composed mainly by methanogenic, sulphate reducing bacteria (SRB) and nitrate reducing bacteria. Furthermore it is reported that co-metabolism is one of the main processes regulating the biodegradation of PAHs by mixed microbial communities. In fact the degradation of a PAH-mixture appears as a co-operative process involving a consortium of strains with complementary capacities. In addition, the presence of more different PAHs can lead to the inhibition or to the stimulation of microbial degradation of other PAHs. For instance Naphtalene was found to inhibit the phenanthrene degradation for a pure culture of *Sphingomonas sp.* (Shuttleworth and Cernaglia, 1996), while was found to stimulate its degradation for a pure culture of *Pseudomonas putida* strain KBM-1 (Haritash et al., 2009). Therefore, the co-metabolism process of PAHs is expected to complicate further the fate of these pollutants in heterogeneous media such as soil, sludge and wastewater.

1.5 Use of the sorption memory of biofilms and polydimethylsiloxane (PDMS) in order to locate pollutants in sewers.

A lot of effort is invested to improve the removal of pollutants by biological processes already existing in the wastewater treatment plants (WWTP), but this must be coupled with a suitable prevention strategy in order to limit the primary introduction of PAHs into the sewers. A problem is to determine the source of pollutants. In this regard, one of the most efficient solutions is the constant monitoring of pollution throughout the sewers and therefore a very cost effective monitoring approach would be required to allow the sewer biofilms to be used as monitoring devices for hydrophobic pollutants due to their natural affinity for these compounds. For a long time microbial biofilms have been studied as bio-monitors for organic and inorganic pollutants in lake, streams and rivers. The direct analysis of the biofilm matrix can provide the identity of the pollutants accumulated into it. Hydrophobic molecules display a very low solubility in water, therefore in aqueous systems they are not homogenously distributed.

This approach exploits the sorption capacity of microbial biofilm for organic and inorganic pollutants. The solvent extraction, the concentration and the detection of the pollutants are the main procedures practiced so far for the detection of pollutants in microbial biofilms (Kostel et al., 1999; Mahfoud et al., 2009, Houhou et al., 2009, Gasperi et al., 2010). An interesting example of this approach is given by Gasperi et al., 2010, a short review focused on the use of sewer biofilms as bio-monitors for tracing the sources of specific pollutants (PAHs and heavy metals) into the combined sewer system of a small catchment area in the centre of Paris, France. Through the analysis of sewer biofilms, it was possible to evaluate the individual contribution of wastewater, runoff and in-sewer processes, to the pollution of PAHs and heavy metals in the Seine river downstream the catchment outfall. This study shows that biofilms might be used for tracing and addressing more precisely different sources of specific pollutants along a sewer.

Although many reported studies from literature are focused on the use of microbial biofilms as bio-monitors for environmental pollutants, most of them refers to relatively small areas, does not make use of a suitable on-field analytical methods and does not involve a systematic sampling strategy.

For these reasons, it would be worth to sample biofilms all along the wastewater distribution system within an urban and industrial area, upstream the WWTP catchment, along all the sewers. This would help to trace sources of organic and inorganic pollutants involving all the catchments connected to the main wastewater

1 Introduction

network (Figure 8). Furthermore, the establishment of a suitable on-field analysis method would improve the potential of this approach, decreasing the time needed for the detection of the pollutants. The analysis of different biofilm matrices sampled along the sewers can provide information about "when" and "where" the pollution started by localization of the contamination in reference to the flow direction. This is a suitable approach point for an effective prevention strategy by identification of the polluter.

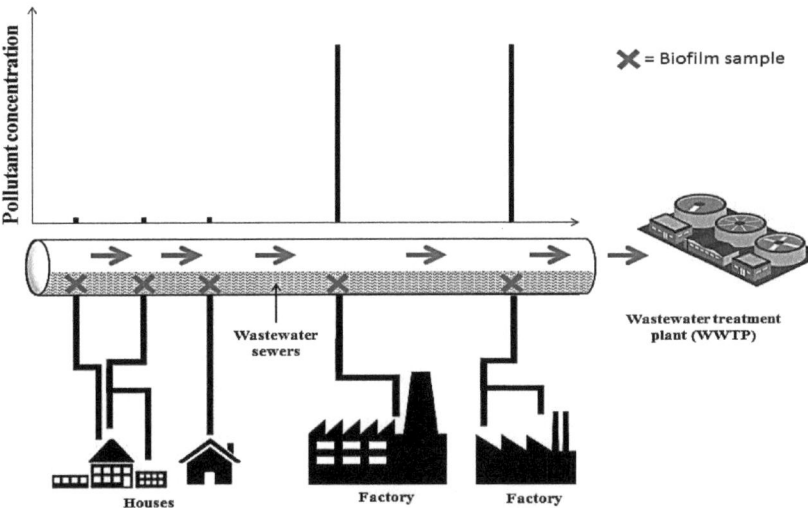

Figure 8 Sketch map of the approach proposed by this study. The measure of the pollutant concentration sequestered by the biofilms along the sewers can help to address the primary source of pollution.

In this regard a really interesting example is shown by Bendt et al., 2007. In this paper is reported the use of sewer biofilms as "sewer spies" for tracing the pollution of heavy metals into the waste water system upstream from the WWTP in the area of Wuppertal, Germany.

For this investigation, the metal load of each biofilm sample was calculated using the atomic absorption spectrometry (AAS) and the geographic information system (GIS, QGIS software) was used to establish the sampling points and facilitate the addressing of the polluters. The biofilms were grown on plastic supports immersed in the wastewater stream and the biomass was scratched off for the analyses.

1 Introduction

This application was proved to display a high potential for detecting also the pollution of fluorinated suractants in sewers and has been further considered for editing new and more accurate environmental policies and regulations in Germany (Genuit and Block 2009).

In order to exploit microbial biofilms for tracing PAHs (phenanthrene) in sewers, it is necessary to estimate the memory of these matrices for such a compound. After the absorption into the biofilm, how long is phenanthrene retained inside the matrix before to be released back into the aqueous phase by desorption processe? This information is necessary to understand whether biofilms can be used as accumulation devices for phenanthrene. A short memory would not facilitate the sampling procedure and all the monitoring approach would be compromised. This manuscript was inspired by the work previously carried out by Antusch and colleagues in 1995 (Antusch et al., 1995), by Koch 2002 and by Genuit and Block in 2009. The hypothesis is that sewer biofilms can accumulate PAH compounds in wastewater and display a long memory for these pollutants.

In order to provide the best possible monitoring procedure, it is necessary to develop an analytical approach which can allow on-field measurements. Although microbial biofilms display sorption capacity for both inorganic and organic pollutants, their matrix remain complex and difficult to analyse by on field measurements. All the methods for the analysis of microbial biofilms need to be performed in a laboratory and depending on the nature of the pollutant, some technical issue migh arise. Furthermore, the biofilm memory for PAH compounds might be too short and not exploitable for the establishment of such a systematic monitoring procedure. There is a substantial lack of information about the memory of microbial biofilms for organic pollutants. More research should be focused on this subject, since microbial biofilms are renewable sorbent materials which can significantly impact the costs of environmental monitoring strategies.

However, when these problems can compromise the all idea, it is auspicable to consider other matrices, which could provide more suitable sorption properties and higher potential for this purpose.

Polydimethylsiloxane (PDMS) is an alternative suitable material for this kind of analysis on PAH pollution in aqueous systems. This oil material is widely used for coating SPME fibers for the extraction of hydrophobic pollutants in untreated environmental samples and displays suitable properties for the employement of fluorescence spectroscopy as method for on field measurements. PDMS is a valid candidate for the creation of monitor devices which might be a good alternative to sewer biofilms.

In order to study the potential of a systematic approach for monitoring the pollution of sewers, this study has been focused on the development of a novel passive approach for detecting phenanthrene in sewers. This part of the work has been attempted both by using sewer biofilms and PDMS oil.

1.6 Polysaccharide gels as samplers for PAH pollutants in water systems.

This investigation was focused on the application of sewer biofilms and PDMS oil for monitoring phenanthrene in sewers. Another aim of this study was to research new possible materials which could be exploited for the same purposes. In this chapter a novel idea will be elaborated in order to stimulate further investigations.

As previously mentioned, the memory of microbial biofilms for hydrophobic pollutants in water has not yet been yet estimated. Microbial biofilms are highly heterogenous matrices and their own properties can largely vary depending on their structure and composition. Also the analysis of biofilms might represent a crucial limitation to the research process. Usually, the most common approaches would involve the optimization of standard protocols of extraction, clean-up and separation of the target compounds in the samples. These procedures are time consuming and can be carried out only by using expensive equipment, such as chromatography instruments (e.g. gas/liquid-chromatography) and special detectors (e.g.mass spectrometry, light spectroscopy). Maybe there is an easier way of achieving the same results by simpler analytical procedures. The main sorption properties of biofilms have been largely investigated and there is a relatively good knowledge on

1 Introduction

the structure and on the composition of the EPS matrix. In this regard, the attention could be focused on the sorption properties of each major class of polymers, which form the EPS matrix. When proved to be suitable sorbent for different pollutants, some of those polymers might be separated and used for developing new sampling devices, without involving the use of the whole biofilm. Polysaccharides are among the most significant consituents which affect the sorption properties of biofilms. If polysaccharide gels exhibit comparable sorption properties to biofilms, then might be more conveninent to use them as sorbent matrices. These polymers are well established industrial products. Natural polymer materials represent a very attractive resource due to their environmental friendly properties. Coming from animals, plants and microorganisms, these materials are abundant in nature, and after being disused, are easily degraded in water into nontoxic derivatives. Furthermore, the production of these materials is fully independent from petroleum resources (Zvezdov and Zvezdova, 2010). Polysaccharide gels have been often employed as models for studying the diffusion of organic chemicals in microbial biofilms, therefore their memory might be comparable to biofilms (Johnson et al., 1996; Jouenne et al., 1994; Cheng et al., 2012; Miller et al., 2004).

It is well known that biofilms can absorb solutes from the water phase, accumulate them and release them back into the aqueous phase, when the environmental conditions change. According to this, also the absorption of organic chemicals in polysaccharide gels may be reversible and needs to be investigated in order to optimize their use as sorbent materials.

1.6.1 Polysaccharide gels and microbial biofilms: sorption of hydrophobic pollutants

Since the purpose of this work is also to provide the reader with reviewed data about passive sampling systems which might be an alternative option to sewer biofilms and PDMS oil, this chapter will attempt to highlight similarities between biofilms and polysaccharide gels. Biofilms have been previously used as sorbent materials for hydrophobic pollutants in water. When polysaccharide gels display properties comparable to biofilms, then their potential can be remarked. During this review microbial biofilms will be treated as a single class of sorbents, but only for facilitating the argumentation. In fact, there is no biofilm which is fully identical to another. Even adjacent portions of the same biofilm sample can exhibit very different physico-chemical properties.

Molecular diffusion is the main mass transport of organic molecules in biofilms and therefore the first part of this chapter will be focused on the diffusion coefficients of different chemicals in both biofilms and polysaccharide gels. In Table 16 are presented the most relevant values of diffusion coefficients, which were previously reported in literature for model molecules. Agar is one of the most abundant polysaccharides produced in industry, therefore was included as reference compound for this class of polymers. As previously mentioned, agar gels have been widely used as surrogates for studying the diffusion of organic chemicals in microbial biofilms (Beyenal et Lewandowsky, 2000; Golmohamadi, 2012; Jouenne et al., 1994; Tresse et al., 1995; Westrin and Axelsson, 1991) and display structural properties, which are similar to those observed in the the EPS matrix. Agar consists of two fractions, agarose and agaropectin (Armisen and Galatas, 1987). Agarose is a long-chain molecule formed by β-D-galactopyranose residues connected through C-1 and C-3 with 3,6-anhydro-L-galactose residues connected through C-2 and C-4 (McHugh, 1987). Both residues are repeated alternately (Armisen and Galatas, 1987). Agaropectin has the same basic disaccharide-repeating units as agarose with some hydroxyl groups of 3,6-anhydro-α-l-galactose residues replacing by sulfoxy or methoxy and pyryvate residues (Fu and Kim, 2010). Agarose and agaropectine togheter form a neutral hydrocolloid presenting polar moieties.

1 Introduction

Table 16 Diffusion coefficients of phenanthrene. Values of D (Diffusivity) for different compounds (left column) into different sorbents (second column from the left) are presented and the references are listed on the right column.

Compound	Sorbent matrix	Diffusivity (D) cm²/s	Reference
phenanthrene	Water	$5-7 \times 10^{-6}$	EPA 1989, Ong et al., 2008
phenanthrene	Biofilm (*Sinorhizobium sp.*)	$2.3 - 4.5 \times 10^{-10}$	Wicke et al.,.2007
α-pinene	water	6.3×10^{-6}	Miller and Allen 2004
Bovine serum albumin (BSA)	Agarose gels 4-7% (v/v)	$1.01 \pm 0.7 \times 10^{-6}$	Johnson et al., 1996
Bovine serum albumin (BSA)	water	$0.5\text{-}0.7 \times 10^{-6}$	Martinsen et al., 1992
Fluorescein isothiocyanate	Water	$5\text{-}6 \times 10^{-6}$	Galambos and Forsters 1998
Fluorescein isothiocyanate	Agar 1.5% (w/v)	4.7×10^{-6}	Wolfaardt et al., 1993
Fluorescein isothiocyanate	Mixed biofilms	7.7×10^{-8}	Lawrence et al., 1994
Toluene	Water	8.6×10^{-6}	
Toluene	Biofilm (*P.putida*)	1.3×10^{-7}	Holden et al., 1997

As reported in the table, the diffusivity of phenanthrene and other molecules into polysaccharide gels is similar to the relative water diffusivity, while in microbial biofilms these values change significantly. Wicke and colleagues reported a diffusivity value of phenanthrene in biofilms of *Sinorhizobium* sp. equal to $2.3 - 4.5 \times 10^{-10}$ cm²/s, while in water is about $(7.0 \pm 0.7) \times 10^{-6}$ cm²/s (Wicke et al., 2007). Between the two values there is a difference of 4 orders of magnitude. Lawrence et al., 1994 reported that the diffusivity of fluorescein (fluorescein isothiocyanate, FITC), measured by fluorescence recovery after photobleaching (FRAP) in agar gel 1.5% (w/v) was found equal to 4.7×10^{-6} cm²/s, while its water diffusivity was reported in Galambos and Forsters 1998 between 5 and 6×10^{-6} cm²/sec. Also for the bovine serum albumine (BSA) a diffusivity of $0.6\text{-}0.7 \times 10^{-6}$ cm²/sec in agarose gels of 4%-7% volume fraction was reported in Johnson et al., 1996. In water the diffusivity of BSA is $0.4\text{-}0.7 \times 10^{-6}$ cm²/sec (Gibizova et al., 2012). In both cases the diffusivity of the target compound in polysaccharide gels with high water content (>90%) was found to be similar to the water diffusivity.

Miller and Allen (Miller and Allen 2004) determined a diffusion coefficient for α-pinene in agar gel 1.5% (w/v) equal to 3.4×10^{-6} cm^2/s, which is comparable to its own water diffusivity. The α-pinene is an organic compound of the terpene class produced by many kind of coniferous trees. It displays a water diffusivity of 6.3×10^{-6} cm^2/s. The diffusivity of fluoresceine (FITC) in mixed biofilms is reported equal to 7.7×10^{-8} cm^2/s (Lawrence et al., 1994). In this case a decrease of two orders of magnitude (98%) is calculated between the diffusivity in water and in biofilms. Holden and co-workers (Holden et al., 1996) reported a diffusivity of 1.3×10^{-7} cm^2/sec for toluene in *Pseudomonas putida* biofilms. The diffusivity of toluene in water is 8.6×10^{-6} cm^2/sec. This is a similar decrease of 98% of the values measured in water, same as reported by Holden et al., 1996. This phenomenon has been explained as hindered diffusivity. According to some authors, both the diffusivity and the hindrance effect for a macromolecule inside a porous matrix is size-dependent (Sharma and Yashonath 2007, Dechadilok and Deen 2006) and increases with the molecular weight and molecular radius of the molecule. In fact, fluorescein isothiocyanate has a molecular radius of about 0.5 nm (Ambati et al., 2000) and a molecular mass of about 380 g/mol, while toluene displays a molecular radius of 0.3-0.4 nm (Van der Bruggen et al., 1999) and a molecular weight of 92 g/mol. The hindrance observed on the diffusivity of toluene in biofilms is similar to the hindrance reported for FITC in biofilms. Phenanthrene displays a radius of 0.4 nm (Gotovac et al., 2007) and a molecular weight of 180 g/mol. Therefore its diffusivity in biofilms might be also lower than its water diffusivity. Therefore it is clear that in biofilms, the diffusion of organic molecules is affected by a strong hindrance, which is not observed in polysaccharide gels.

Now another aspect of the sorption mechanism must be taken into account. The mass transfer of molecules in biofilms it is mainly, but not only driven by molecular diffusion inside the aqueous phase. The water mobility is an important parameter to be investigated. Although in both matrices, polysaccharide gels and biofilms, the mobility of water is mostly similar to its mobility in free water, further specifications are to be made. Beuling and co-workers (Beuling et al., 1998) used pulsed field gradient nuclear magnetic resonance (PFG-NMR) in order to measure the diffusivity of water inside microbial biofilms and agar gels. They observed that the diffusion coefficient of water in agar gels of 1.5% (w/v) - 2% (w/v) ranges between 90%-95% of the diffusivity in pure water (20 x 10^{-6} cm^2/sec) linearly decreasing with increasing the gel concentration. Therefore in pure polysaccharide gels the mobility of water does not seem to be affected significantly. But when 20%-30% (v/v) of polystyrene particles (0.9 µm diameter) and bacterial cells are added into two separate agar gels (1.5% w/v), the water diffusivity decreases similarly to 85% of its value in pure water. These two kinds of gels display same hindrance on the main fraction of water. From this, the authors conclude that the cells seem to affect the mobility of water mostly by reduction of the diffusive volume and by steric interactions. However, from the PFG-NMR spectra obtained analysing the polysaccharide gels enriched with bacterial cells, it was possible to detect at least 2 different water fractions characterized from two different diffusivities. These fractions where not found in the agar gel enriched with polystyrene particles.

From those results it was reported that the main portion of the water displayed similar diffusivity as observed in the other matrices, but the smallest water fraction exhibited a lower diffusivity, which the authors define as intracellular water. In gels enriched with bacterial cells, with a cell fraction of 5% v/v, the diffusivity of water is 90%-95% of its self-diffusivity in pure water. This difference decreases linearly with the cell volume fraction. In 2000 Vogt and colleagues reported new data about the water mobility in microbial biofilms (Vogt et al., 2000). In biofilms of *Ps.aeruginosa* after 24 hours of growth with a cell fraction of 1% v/v, the total water volume is divided in at least three fractions, each one characterized by a different self-diffusivity. The main portion of water displays a diffusivity equal to 85% of its value in pure water. This is the so called "free water fraction" in the biofilm. The second fraction (0.4% v/v) displays a diffusivity that is 10% of the pure water self-diffusivity. This is defined as intracellular water fraction. The third water fraction (<0.1% v/v) displays a self-diffusivity which is 1% of the value in pure water. This water is the water entrapped in the secondary structure of the EPS polymer matrix. In addition to this, within the same study, the diffusivity of glycerol was investigated in biofilms of *Ps. aeruginosa*. For glycerol two main portions were identified: the main fraction (90%) with a diffusivity of 80%-85% the diffusivity of glycerol in pure water, which is assigned to the glycerol in the pores of the biofilm filled with water. The second portion (10%) displays a value of diffusivity equal to the 1% of that in pure water. This is assumed by the authors to be the glycerol diffusing in the EPS network and partially entrapped in the secondary structure of the polymer matrix. According to these results, the mass transfer of solutes in a biofilm occurs as diffusion and the main fraction of a solute (glycerol) in such a matrix displays diffusivity values equal to 80%-90 % of its diffusivity in pure water and is addressed in the water filled pores of the matrix. A smaller fraction of the solute (10%) though is entrapped in the thiner EPS networks and its diffusivity is significantly reduced. Such a decrease (99%) in the EPS was reported for only 1% of the water fraction. Glycerol might display much stronger ionic interactions with the polymers in of the EPS matrix than a nonpolar compound such as phenanthrene. For phenanthrene the main decrease in the diffusivity through the EPS matrix is expected to be due to steric interactions, tortuosity and weak electrostatic interactions, which are the same causes of diffusivity hindrance of water in polysaccharide gels.

Two main types of water have been defined in biofilms: free water and bound water (Beuling, 1998; Vogt et al., 2000). The free water is the most abundant type of water, while bound water represents the least fraction (1%). Beuling 1998 reported a self-diffusivity of water in biofilms equal to 11×10^{-6} cm^2/sec, compared with a pure water self-diffusivity of 20×10^{-6} cm^2/sec (between 20° C and 30°C). A fraction of 10 % of the water displayed a diffusivity of 19×10^{-6} cm^2/sec. The significantly decreased water self-diffusivity refers to about 1% of the all water volume and it is mostly located in the intracellular portion of the biofilm. Therefore, polysaccharide gels with immobilized bacterial cells display two different water phases: outside and

1 Introduction

inside the cells. The diffusivity of solutes in the cells can be significantly reduced (1 order of magnitude according to Vogt et al., 2000). Increasing the concentration of cells in a biofilms, the total diffusivity of a solute inside a biofilm decreases linearly (Beuling et al., 1998).

Cells and secondary structures of polymers are the main components responsible for the strongest hindrance of water diffusivity in biofilms. EPS matrix and polysaccharide gels display the same predominant water mobility and this reveals similar properties regarding the diffusion of macromolecules such as phenanthene (Vogt et al.; 2000). Biofilms are heterogeneous matrices composed of different constituents with different physical-chemical properties (Table 30). The diffusion of a solute changes depending on which one of this phases interacts with it.

It seems reasonable to say that the mass transfer of macromolecules in the EPS matrix is not well modeled by polysaccharide gels. In the EPS matrix, a portion of the molecules (minor) might exhibit a very lower mobility. When this minor fraction of the target molecule is still detectable by very sensitive analytical techniques, then the memory of the biofilm might result significantly longer than the memory of a polysaccharide gel.

Not more than these speculation can be carried out from the reviewed data, because the memory of polysaccharide gels for phenanthrene has never been empirically measured. However, when these speculation would match experimental results, polysaccharide gels may not be suitable matrices for replacing microbial biofilms as sorbent materials. A solution might be to add in the device, together with the polysaccharide gel, a small volume of a sorbent material, which displays a very high partition coefficient for the target molecule in water. In such a case, this device would rapresent a suitable alternative to the microbial biofilms. A similar idea was presented by Chen and colleagues, who manufactured a device composed of an outer layer of agarose gel and an inner phase filled with AMBERLITE™XAD18, a special polymeric adsorbent resin (Chen et al., 2012) used for detecting antibiotics in surface water. In such a case, polysaccharide has been efficiently used.

1.7 Aims of this study

The aims of this study were:

- Choose a reference PAH compound based on the environmental occurrence and its chemical properties.
- Choose the analytical method for the experimental part of the work.
- Investigate the properties of polysaccharide gels for the development of passive sampling devices.
- Choose a suitable method in order to collect biofilm samples from wastewater streams.
- Design and construction of all the additional equipment needed for analyzing the phenanthrene into sewer biofilm suspensions using front-face fluorescence spectroscopy.
- Research of an alternative bio-monitor material rather than sewer biofilms and polysaccharide gels for monitoring PAHs in wastewater streams and test it using the analytical method previously selected for the sewer biofilm analysis.
- Suggestion for monitoring devices

2 MATERIALS AND METHODS

2.1 Materials

2.1.1 Polydimethylsiloxane (PDMS) monitor device

A self-designed device filled with PDMS oil (Wolfgang Spielberger eK) was used for sorption experiments in deionized water and in a pond water suspension. The device is composed of three main stainless steel components: two external sealing frames and one internal chamber with a volume capacity of 40 mL. Two dialysis membranes foils (SERVA Visking dialysis membrane, 12-14 KDa porosity) were placed between the stainless steel sealing frames and the inner chamber as shown in Figure9a. In order to ensure the airtightness of the seal, two metal frames and two additional rubber frames, made of ethylene propylene diene-monomer rubber (EPDM from Schmidt & Bartl GmbH, 65 shore), were placed between the membrane foil and the internal chamber (Figure 9c). The dialysis membranes were washed for an hour in deionized water prior to use. The external stainless steel frames were then sealed with stainless steel screws. Once closed, the device was filled with PDMS oil. The total volume capacity of the device is 40 mL and the size is 5 cm x 4 cm x 3 cm (length x height x thickness).

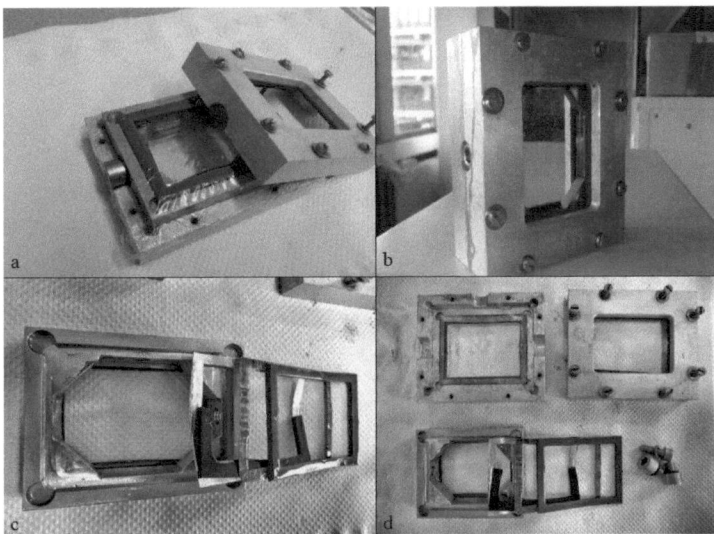

Figure 9 Prototype PDMS monitor device. a) external stainless steel frames and whole system with the dialysis membranes. b) assembled device ready to be tested. c-d) the different sealing frames which placed between the inner chamber and the dialysis membrane; the first frame is made of tin (99%) the second is made of EPDM, the third is a copper frame and one more EPDM (black and grey color) d) a top view of all the components of the device before to be assembled.

2.1.2 Optical fiber platform

A self-designed fiber optic platform was manufactured at the University of Duisburg-Essen (Figure 10). This device is made of aluminum and is composed of two different planes. The optical fiber support can be moved at different height levels by a worm shaft mechanism. The sample plate can be moved on the x-axel also by a worm shaft mechanism. The angle between the two fiber optic collimators is fixed at 120° while the sample can be rotated in order to reach the better angle of incidence for the excitation of the sample surface (Figure 10c). The base has size 12 cm x 1.5 cm x 5.5 cm (length x height x thickness). The worm-shaft support on the bottom of the platform base, on the z-axe has size of 6 cm x 2.5 cm x 1 cm (length x height x thickness).

2 Materials and methods

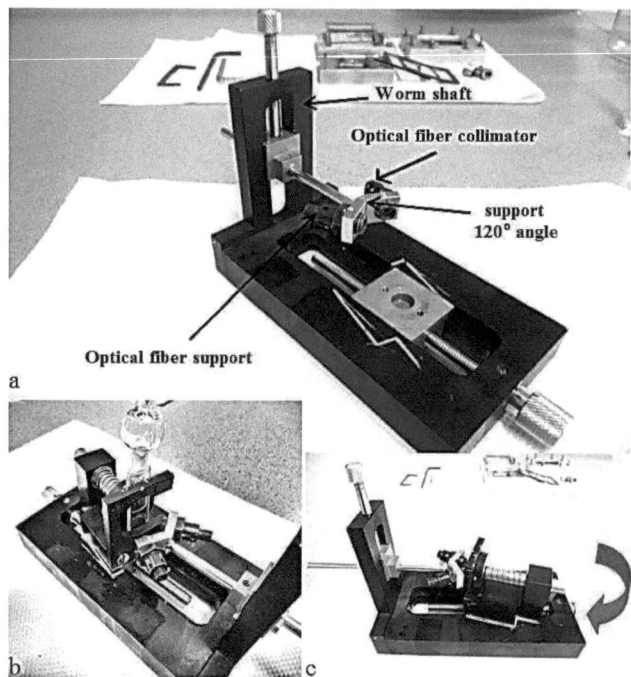

Figure 10 The holder for the optical fiber devices. a) the angle of the optical fiber collimator is set at 120°. The optical fiber support, the collimators and the worm shaft mechanism are indicated with the black arrows. b) the sample is placed in front of the optical fiber shaft. c) the sample holder can be moved on the horizontal plane (red arrow) in order to achieve a better angle of incidence for the excitation of the sample. In this example is shown a cuvette holder, but this platform was used for the analysis of the PDMS prototype device. The sample presented in the image is a 500 µL quartz cuvette.

2.2 Analytical Instruments

2.2.1 Bench fluorescence spectrometer

For the Uv-fluorescence calibration and measurement with 3.5 mL QS glass cuvettes a Shimadzu RF-5301PC Shimadzu fluorescence spectrometer equipped with a xenon arc lamp was used. The excitation wavelength was set at 290 nm.

2.2.2 USB fluorescence spectrograph

The fluorescence measurements on the PDMS device were performed with Ocean Optics USB QE65000 fiber optic USB spectrometer (Figure 11). Spectrasuit, a java based spectroscopy software, was exploited for the acquisition of the emission signal and the parameters were set as following (Table 17):

2 Materials and methods

Table 17 Software parameters.

Sample	Integration time	Average scans per sec.	Boxcar width
Sewer biofilm	30sec	2 scans	4 nm
PDMS experiments	2 sec	2 scans	4 nm

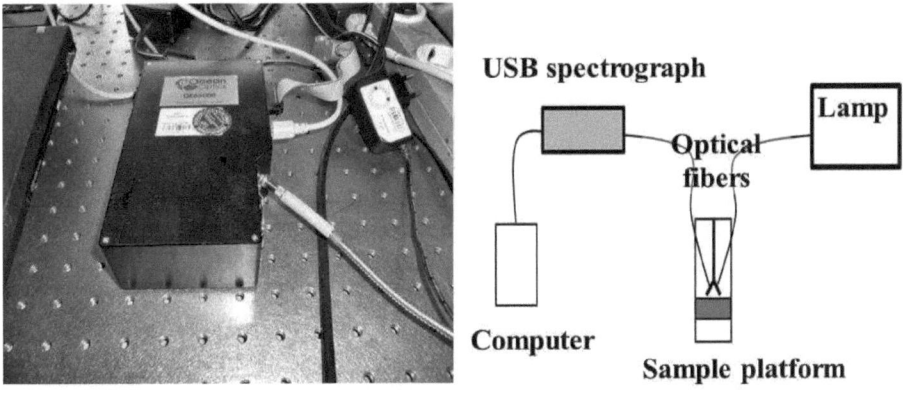

Figure 11 Ocean optics USB spectrometer and a sketch of the analytical system used for detecting phenanthrene in sewer biofilm suspensions and PDMS oil.

2.2.4 GC-MS device

The extraction of phenanthrene from the biofilm samples was performed according to the DIN38414-23 method. The GC-MS SIM mode analysis was performed according to Qian et al.,2011 (GCMS-QP5050, Shimadzu, Duisburg, Germany) under the following conditions: HP-5MS capillary column (0.25 mm , 30 m, 0.25 lm thick), helium 5.0 as carrier gas, flow rate at 1 mL/min, injector temperature at 250° C, interface temperature at 300° C, SIM mode, and electron ionisation (EI). The temperature program started at 160° C, held for 2 min and increased to 280 °C at a rate of 10° C per minute. After reaching 280° C, the temperature was held for 10 min.

2.3 Methods

2.3.1 Preparation of aqueous solutions of phenanthrene

2.3.1.1 PDMS device experiment in deionized water solution

A methanol solution (HPLC grade from Fisher scientific UK) of phenanthrene at 100 mg/L was prepared inside a 10 mL matrass flask (Blau brand duran glass) diluting a methanol stock solution of phenanthrene at 1000 mg/L. The methanol solution of phenanthrene at 100 mg/L of concentration was stirred for ten minutes and then mixed with 1 L of deionized water solution (previously autoclaved) in order to reach a phenanthrene solution concentration of 1.5 mg/L. The whole preparation was performed under sterile conditions. This experiment was carried out three times under the same conditions.

2.3.1.2 PDMS device experiment in pond water solution

A volume of 2 L of water was sampled from a pond reservoir located in the city of Essen, Germany. The sample solution was divided in two volumes of 1 L and each volume was spiked with 1.5 mL of a methanol stock solution (HPLC grade from Fisher scientific UK) of phenanthrene at 1000 mg/L. Both solutions were stirred for 30 minutes with a magnetic stirrer (800 rpm). One solution was used for the absorption of phenanthrene in the PDMS device and the other solution was used as a blank in order to monitor the decrease of phenanthrene during the experiment without the use of the device.

2.4 Sampling of the sewer biofilm

The biofilm samples were collected from the Bielefeld sewer system using a special plastic support (Figure 12a) kept into the wastewater stream for 4 weeks. After this period, the sampler was taken out from the sewer and the biofilm was scratched (Figure 13a) and collected in a sampling bottle. The samplers were placed along various sites of the sewer system in Bielefeld city, downstream to different wastewater sources (Figure 12b, Table 18).

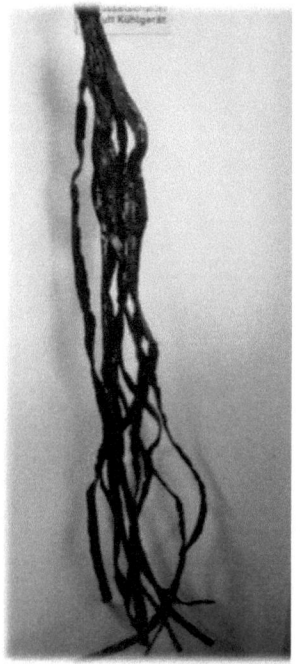
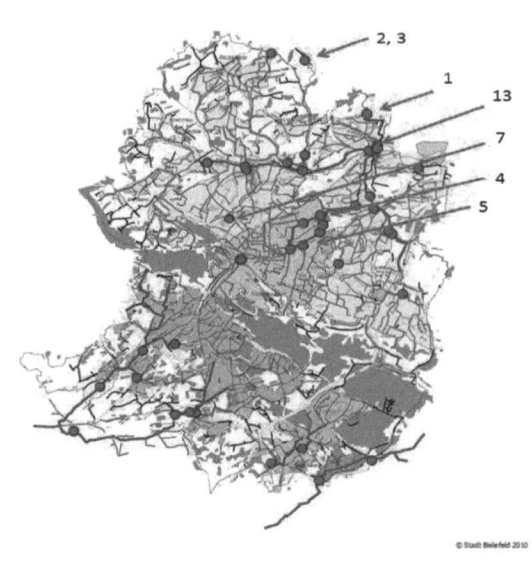

Figure 12 Overview of the sampling sites within the Bielefeld sewer system. The biofilm samples analyzed by GC-MS are indicated by the red arrows. On the left is shown the plastic sampler device used for collecting the biofilm from the sewers in collaboration with the Umweltamt of the Bielefeld city hall administration

2 Materials and methods

Table 18 Samples collected from the Bielefeld sewers by the Umweltamt city administration department. For each sample are indicated: the number, the name, the date of sampling and the specific upstream source of wastewater. Different sources might lead to a different composition of the biofilm matrix.

Number	Name	Date	Upstream wastewater source
1	419uth	7/8/2013	Metal (powder coating)
2	417EJ-DB	7/8/2013	Disposal site
3	417EJ-DS	7/8/2013	Disposal site
4	425hDrew	7/8/2013	Metal (powder coating + electroplating)
5	103WarG	7/8/2013	Metal (anodized) + screen printing
6	104nLWN	7/8/2013	Metal (Rinse + degreasing)
7	413GH	7/8/2013	Dentist
8	529JVA	7/8/2013	Laundry, Great kitchen
9	530nBAU	7/8/2013	Iron Foundry
10	424WS	13/8/2013	Mixed Municipal wastewater
11	425SP	13/8/2013	Mixed Municipal wastewater
12	419KA	13/8/2013	Mixed Municipal wastewater
13	400HN	13/8/2013	Mixed Municipal wastewater

Figure 13 Overview of the devices used for sampling sewer biofilms. a, b) the biofilm is scratched off from the sampler into a special tool. c) a closer look at the sewer biofilm attached on the PFT support used for the sampling procedure.

2.5 Front face fluorescence measurements on the PDMS device

The PDMS prototype device was placed on the fiber optic platform, in order to achieve an angle of incidence equal to 30° for the excitation of the PDMS surface. This angle was chosen as the best arrangement in order to guarantee the reproducibility of the measurements (Figure 14).

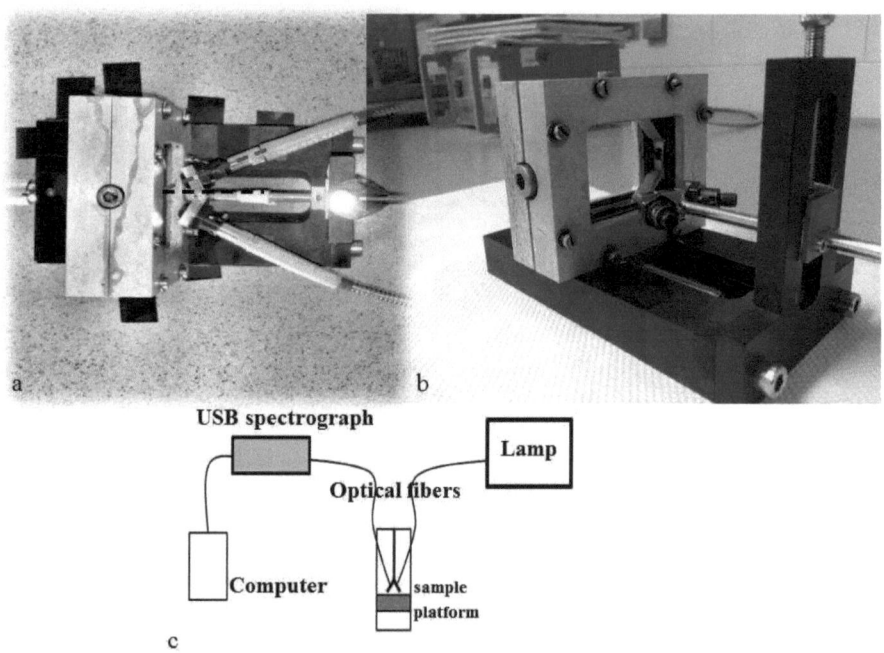

Figure 14 Top view of the analytical system arranged for the detection of phenanthrene inside the PDMS prototype device. a) The angle of incidence between the perpendicular to the target surface (black line) and the light beam (red line) was set at 30°. b) a closer view of the whole system. c) sketch of the experimental system.

2.6 Calibration procedures

2.6.1 The quantification of phenanthrene in deionized water solution.

A stock solution of 1000 mg/L of phenanthrene was prepared mixing 100 mg of phenanthrene crystals (Supelco Analytical, 99.1% purity) in 100 mL of pure methanol (HPLC grade Fischer scientific UK). The solution was stirred for 15 minutes by magnetic bar at 800 rpm. Then a methanol solution of phenanthrene at 10 mg/L was prepared. This solution of phenanthrene was mixed with deionized water in order to prepare the standard solutions for the calibration procedure. The following phenanthrene standard solutions were prepared: 1.5 mg/L, 1 mg/L, 0.5 mg/L and 0.250 mg/L. The calibration was performed five times and each standard prepared in triplicate and analyzed in 3.5 mL QS glass cuvettes by using a RF-5301PC Shimadzu fluorescence spectrometer. The limit of quantification and the limit of detection were measured by adding respectively ten times and three times the standard deviation to the value of fluorescence emission intensity obtained from a blank solution of pure deionized water. Due to the low solubility of phenanthrene in deionized water, the maximum concentration measured was 1.5 mg/L and the minimum was 0.25 mg/L. The limit of detection (LOD) and the limit of quantification (LOQ) were 0.05 mg/L and 0.1 mg/L respectively.

2.6.2 GC-MS device

A stock solution of phenanthrene at 1000 ppm was prepared in methanol and diluted 10 times with acetone in a 10 mL flask. From this solution were prepared six standard solutions of phenanthrene in acetone: 1.5mg/L, 1.0 mg/L, 0.75 mg/L, 0.5 mg/L, 0.25 mg/L, and 0.125 mg/L. The standards were prepared in triplicate and analyzed using the previously described method. The area of the peaks was taken as reference for the calibration. The calibration curve is shown in Figure 15.

The limit of detection is equal to 1 µg/L and the limit of quantification is equal to 3 µg/L.

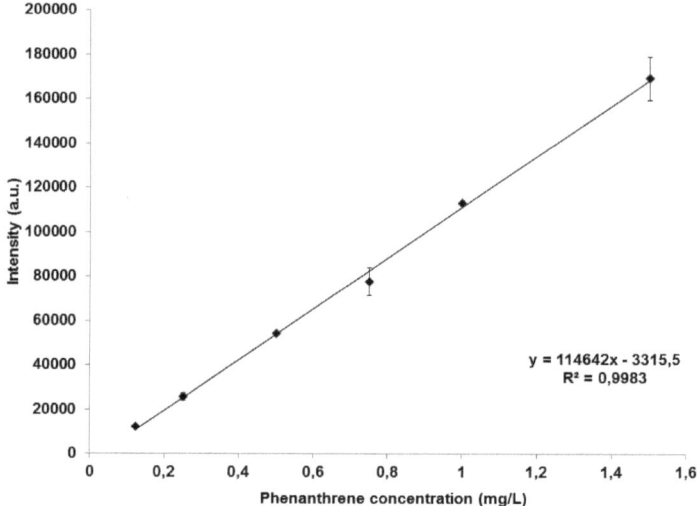

Figure 15 GC-MS calibration curve for the quantification of phenanthrene from sewer biofilm suspensions. The standard concentrations are 0.125 mg/L, 0.25 mg/L, 0.50 mg/L, 0.75 mg/L, 1.0 mg/L, 1.5 mg/L.

2.7 Experiments

2.7.1 PDMS device in deionized water with phenanthrene

A volume of 40 mL of PDMS oil (WS Silikonöl V1000) was poured into the device that was plunged into deionized water solution (800mL) with phenanthrene (1.4 ±0.2 mg/L). The fluorescence emission spectra of phenanthrene absorbed by the device were collected each hour for the first nine hours and then each 24 hours till 72 hours. In order to perform the analysis, the device was removed from the water solution and left 10 minutes in the dark. In this time the semipermeable membrane, which encloses the PDMS oil, was getting dry and straight on its surface, so that allowed to be analyzed by front face fluorescence spectroscopy. After ten minutes the device was placed on the platform where optical fiber devices were used to perform the measurements.

2.7.2 PDMS device in pond water with phenanthrene

A volume of 40 mL of PDMS oil (WS Silikonöl V1000) was poured into the device that was plunged into a 1.4 mg/L pond water solution (800mL) of phenanthrene. The fluorescence emission spectra of phenanthrene absorbed by the device were collected after one hour and after 24 hours. In order to perform the analysis, the device was removed from the water solution and left 10 minutes in the dark. In this time the semipermeable membrane, which encloses the PDMS oil, was getting dry and straight on its surface, so that allowed to be analyzed by front face fluorescence spectroscopy. After ten minutes the device was placed on the optical fiber platform where optical fiber devices were used to perform the measurements.

2.8 Data elaboration procedures

2.8.1 Simulating diffusion of phenanthrene from the self-designed PDMS device.

The following diffusion mass transfer formula was exploited for calculating the theorethical desorption of phenanthrene from the PDMS oil device into the flowing water phase. The simulation was formulated assuming deionized water to be the flowing aqueous phase.

$$\left[\frac{g}{m^2s}\right] = D * A * \frac{Cr - Cl}{x} \tag{Eq.9}$$

where C_r and C_l are the phenanthrene concentration (in g/m^3) in the device and in the water respectively. A is the surface area of the PDMS phase, which is in contact with the water phase; x is the thickness of the PDMS chamber.

3 RESULTS

This study is focused on discussing the use of sewer biofilms for monitoring phenanthrene in sewers. A field trial experiement was carried out adopting a systematic monitoring strategy of the sewer system in the city of Bielefeld. Sewer biofilms were sampled along a section of the pipeline and the analysis of the samples was done by extraction, separation and mass spectrometry analysis. In addition, a new self-designed monitoring device was developed by using commercially available PDMS oil. A novel analytical method was tested in order to investigate the potential of the delf-designed device for running in situ measurements by UV-fluorescence spectroscopy.

3.1 Quantification of phenanthrene in sewer biofilms by GC-MS

No significant amount of phenanthrene was detected inside the biofilm samples taken from the monitored are of the sewers in the Bielefeld city (data not shown).

Further discussion on this outcome will be presented in the discussion section.

3.2 PDMS Device as pollutant sampler and analytical device

Although several attempts were made using sewer biofilm samples, it was not possible to study the sorption kinetics of phenanthrene by using simple steady state fluorescence spectroscopy. Since this technical approach it was considered as the easiest solution in order to carry out measurements by using portable equipment for monitoring phenanthrene in sewers, it became necessary to find alternative and more suitable sorbent materials rather than biofilms. It was necessary to find a UV transparent substance that could display significant accumulation of hydrophobic compounds such as PAHs and that could be analyzed by the analytical equipment designed and manufactured during this work. Polydimethylsiloxane (PDMS) displays these exact properties and therefore it was considered a suitable alternative to sewer biofilms and polysaccharide gels for the sorption and the detection of phenanthrene in sewers. The steady state fluorescence spectroscopy in front face mode is the analytical method of choice. In this part of the work, a prototype device was manufactured and exposed to aqueous solutions spiked with phenanthrene. During the experiment, the device was removed from the solution and analyzed by front face fluorescence spectroscopy equipment. The aim of this preliminary experiment was to test the PDMS device in detecting phenanthrene both from deionized water and from a natural pond suspension. The hypothesis was that the steady state fluorescence spectroscopy is a suitable analytical method for the detection of phenanthrene in wastewater using PDMS oil as sorbent material. At the end of these experiments the PDMS oil contained in the device and exposed to the contaminated aqueous solutions was analyzed using a commercially available bench spectrometer (RF-5301PC Shimadzu fluorescence spectrometer) and the resulting spectra were compared with a solution of phenanthrene at 1.5 mg/L (Fig.17). This comparison procedure was carried out in order to verify the presence of phenanthrene inside the PDMS oil. From the results (Figure 17) it seems that phenanthrene is present inside the PDMS oil after the exposure (24 hours) to the contaminated solutions. In the spectra of the PDMS oil the baseline is significantly higher than the baseline observed in the spectrum of the standard solution of phenanthrene. This might be due to interferences caused by specific components of the EPDM sealing frames, which are in contact with the oil and by other molecules present in the pond suspension (e.g. humic susbtances).

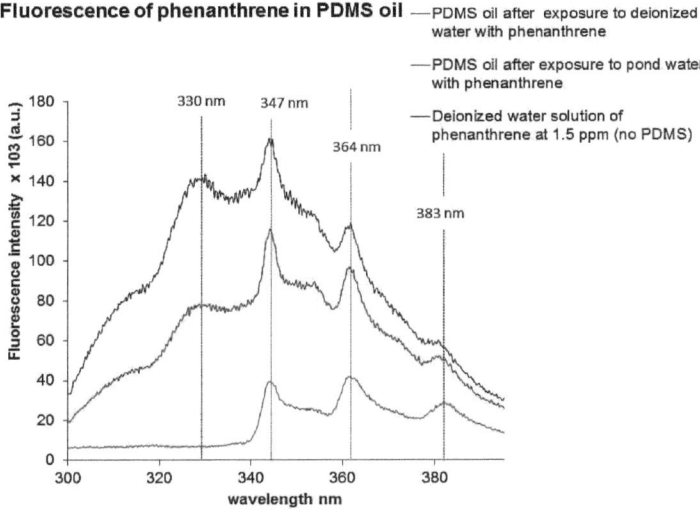

Figure 16 Phenanthrene fluorescence emission spectra obtained by standard procedure of fluorescence measurement in front face mode using a RF-5301PC Shimadzu fluorescence spectrometer. The spectrum of the phenanthrene standard (1.2ppm) is shown as a blue line, In the figure are shown also the PDMS oil spectra after the exposure to the pond water (red line) and to the deionized water (black line) spicked with phenanthrene. It is possible to identify the phenanthrene peaks in these latter spectra.

The analysis of the PDMS by standard procedure revealed the presence of phenanthrene, but the interference due to some component of the rubber sealing frames caused a decrease in the spectroscopic resolution (330 nm). As previously mentioned, the device was designed not only for the accumulation of phenanthrene into the PDMS oil, but also for performing front face measurements in order to identify phenanthrene using self-made analytical equipment. The comparison between the spectra obtained by using a commercially available spectrometer (Figure 17) and the self-designed equipment (Figures 18, 19 and 20) was necessary in order to evaluate the efficiency of the device. In order to facilitate the comparison between the various fluorescence spectra, the peaks of intensity which were equal are marked by the green icon (v) and the differences are highlighted by the red icon (x).

3 Results

For each experiment, in deionized water and in pond water, the device was analyzed before the exposure, after 1 hour and after 24 hours of exposure to the phenanthrene solution and to the blank solution of only deionized water. In Figure 18 are shown the PDMS spectra obtained during the experiment in pond water and in Figure 19 the spectra from the experiment in deionized water.

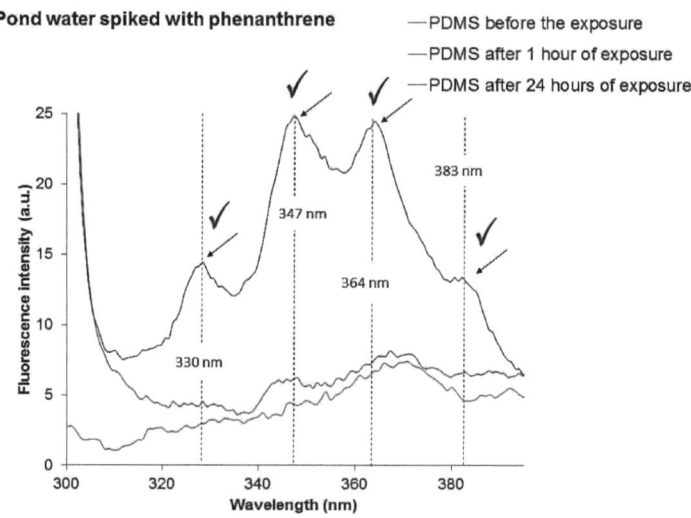

Figure 17 Fluorescence spectra of PDMS oil obtained exploiting the self-designed analytical equipment from the exposure of the device to the pond water spiked with phenanthrene. The green line is the spectrum before the exposure to phenanthrene; the red line is the spectrum after 1 hour of exposure and the black line is the spectrum after 24 hours of exposure. The dot lines indicate the four fluorescence peaks of the PDMS oil spectrum which were previously observed using the standard analytical procedure: 330 nm, 347 nm, 364 nm and 383 nm. The green icon (v) indicate the corresponding peaks observed both in the the spectra obtained by standard and in those from the self-designed equipment.

3 Results

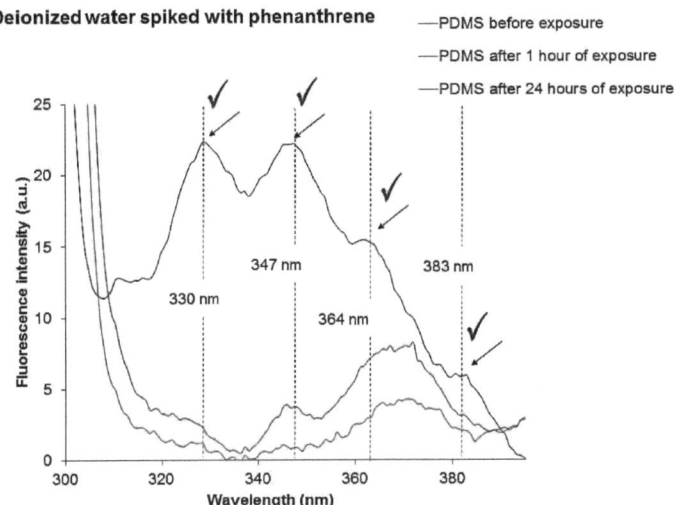

Figure 18 Fluorescence spectra of PDMS oil obtained exploiting the self-designed analytical equipment from the exposure of the device to the deionized water spiked with phenanthrene. The green line is the spectrum before the exposure to phenanthrene; the red line is the spectrum after 1 hour of exposure and the black line is the spectrum after 24 hours of exposure. The dot lines indicate the four fluorescence peaks of the PDMS oil spectrum which were previously observed using the standard analytical procedure: 330 nm, 347 nm, 364 nm and 383 nm. The green icon (v) indicate the corresponding peaks observed in both the spectra obtained by standard and in those from the self-designed equipment.

The same peaks observed in Figure 17 are observed in Figure 18 and 19. In the case of Figure 19 the baseline is higher than in Figure 18 and this is also observed in Figure 17. The device used for all the experiment was the same. The experiments in deionized water were the first ones carried out, therefore it might be that in the PDMS oil, the amount of fluorescent EPDM component was more than the amount present in the PDMS oil after the exposure to the pond water. This is the reason why the spectrum in Figure 18 shows higher resolution of the phenanthrene signal than the spectrum in Figure 19. This is also confirmed in Figure 17. In order investigate the origin of the peak at 330 nm, a blank experiment was carried out in pure deionized water without phenanthrene and the spectra are shown in Figure 20. In this case only the peak at 330 nm is visible, while all the other peaks observed during the exposure to phenanthrene don't rise. This blank experiment was carried out using the same device and the peak at 330 nm displays similar intensity of the corresponding peak in

Figure 18 but lower than in Figure 19. As previously said, most of the amount of the fluorescent compound contained into the EPDM sealing frames was extracted into the PDMS oil during the first experiments in deionized water, therefore during the experiments carried out in a second time (Figure 18 and 20) the peak at 330 nm displays a lower intensity.

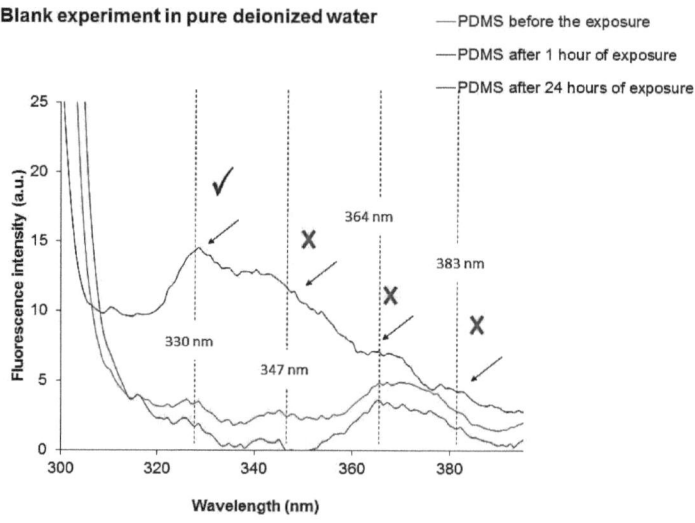

Figure 19 Fluorescence spectra of PDMS oil obtained exploiting the self-designed analytical equipment from the exposure of the device to the pure deionized water without. The green line is the spectrum before the exposure to phenanthrene; the red line is the spectrum after 1 hour of exposure and the black line is the spectrum after 24 hours of exposure. The dot lines indicate the four fluorescence peaks of the PDMS oil spectrum which were previously observed using the standard analytical procedure: 330 nm, 347 nm, 364 nm and 383 nm. The green icon (v) indicates the similarities observed between the spectra obtained by standard and self-designed equipment and the red icon (x) indicates the peaks, which were not observed in this case.

Furthermore, from Figures 18 and 19 it is possible to observe that after 1 hour of exposure, the first fluorescence peak of phenanthrene rose up at 347 nm. This peak was present only when phenanthrene was dissolved in the solution. In the blank experiment this peak does not appear in any one of the performed analyses (after 1 hour and after 24 hours). The similarities observed between the spectra obtained by the standard procedure and by the self-designed equipment, reveal a good level of resolution, although the interference observed in all the spectra affect the unequivocal identification of phenanthrene.

3 Results

Once the device was emptied, the inner volume of the chamber was washed with deionized water and the fluorescence spectrum of the empty device was collected in order to check if phenanthrene was adsorbed on the external surface of dialysis membrane (Figure 21). The aim of this analysis was to confirm that the phenanthrene was absorbed all by the PDMS oil only. It was not possible to detect any of the fluorescence peak characteristic of phenanthrene in the spectrum of the dialysis membrane after the experiment. From the comparison of all the spectra obtained, it seems that the prototype device absorbed phenanthrene from the water phase only into the PDMS oil and not on the membrane. In fact, the spectrum of the membrane shows the only broad peak of cellulose between 360 nm and 380nm.

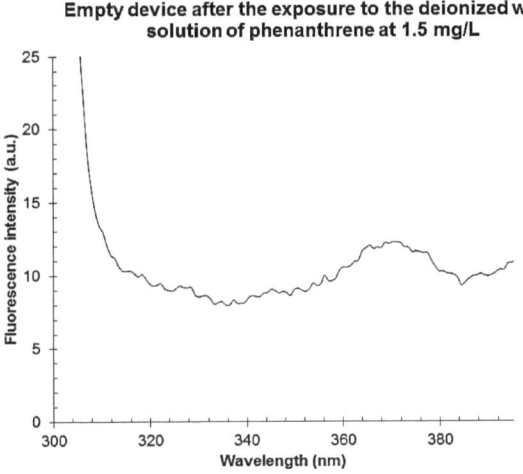

Figure 20 Fluorescence spectrum of the empty device. After the exposure to the deionized water containing phenanthrene at 1.5 mg/L of concentration. The broad peak of fluorescence intensity is the charachteristic peak of the membrane and similar spectra were collected from the device at the beginning of each experiment carried out in this part of the work.

In this regard, must be pointed out that all the spectra shown in the figures display a broad peak of fluorescence around 365 nm, which is the fluorescence spectrum of cellulose polymers when excited by a UV light at 290 nm (Plitt and Toner 1961). The dialysis membrane is made of cellulose, therefore this peak is present in all the

3 Results

measurements and is the only peak observed in Figure 41, when the membrane is analyzed without PDMS oil and phenanthrene.

The next part of the results is focused on the data obtained from the exposure of the device to the dionized water solution of phenanthrene at 1.5 mg/L (Figure 19). During the experiments the system was not hermetically isolated from the atmospheric phase and phenanthrene was subjected to both sorption into the device and evaporation in the gas phase. In real conditions this two processes would be influencing the detection of phenanthrene by the monitor device. The evaporation of phenanthrene in the gas phase does not affect the absorption into the prototype device.

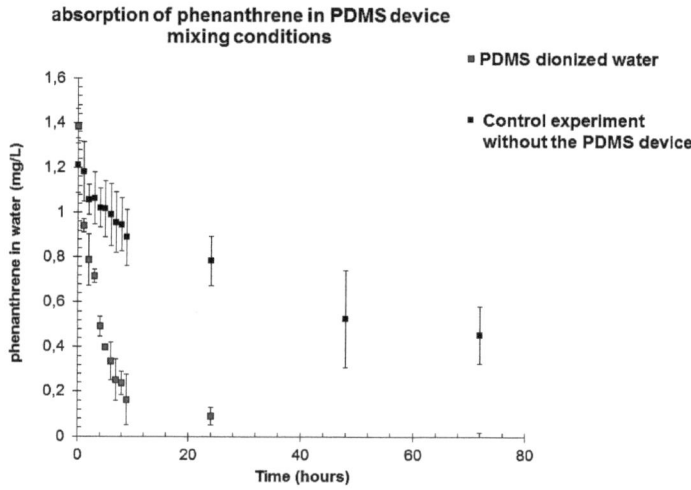

Figure 21 Phenanthrene concentration in water during the test with the device and during the control experiment. When the device is submerged into the dionized water solution of phenanthrene (red line), the decrease of phenanthrene is faster compared with the control experiment (black line). For the control experiment the deionizd water solution of phenanthrene was not in contact with the PDMS device. In both cases the solution was stirred at 900 rpm.

In fact, from Figure 22 it is possible to observe that the absorption process takes place mostly during the first 9 hours and overcomes the evaporation process. After 9 hours the device is saturated and the further loss of phenanthrene follows similar kinetics with or without the presence of the device. Presumably at this point the main process taking place is the evaporation of phenanthrene in the atmospheric gaseous phase. Using a linear fitting curve it is possible to observe that the decay costant of the phenanthrene concentration in the water phase after 8 hours has a comparable value between the two experiments (Figure 24). On the other hand during the first 8 hours the decay constants differ greatly between the two experiments, showing a much faster decrease of phenanthrene concentration when the device is submerged in the water solution (Fig.23).

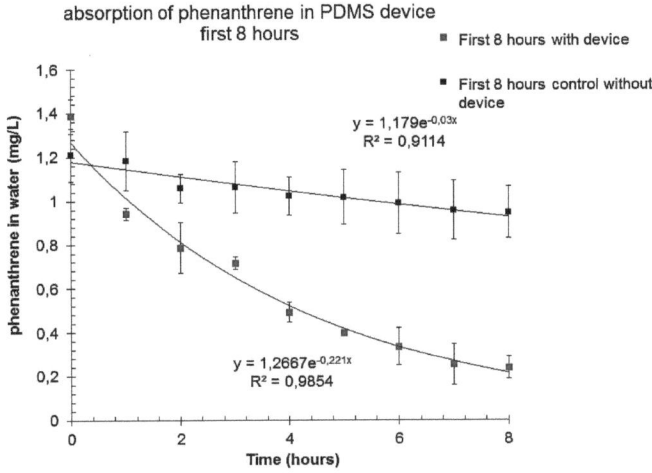

Figure 22 The phenanthrene concentration (mg/L) in the water during both the experiments: with (red line) and without (black line) PDMS device submerged in the solution. In the chart are shown the exponential equations for both decays, in order to compare the different rates. The decay rate is higher when the PDMS device is plunged into the phenanthrene solution.

3 Results

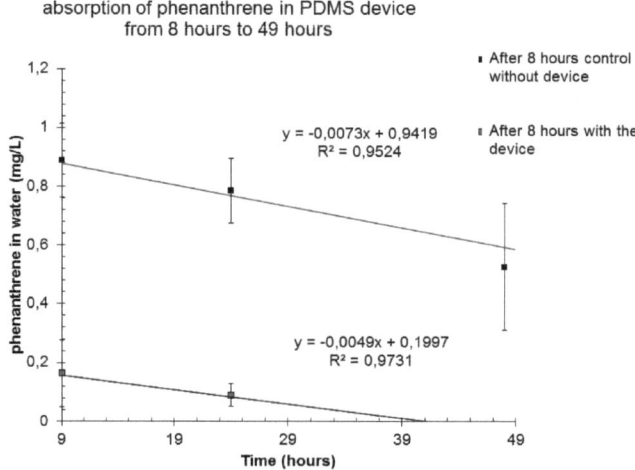

Figure 23 The phenanthrene concentration in the water during the second part of both experiments: with (red line) and without (black line) PDMS device present inside the phenanthrene solution. After the first 10 hours the decrease of phenanthrene in both solutions seems proceeding with similar rates, as shown by the equations

These results show that the use of PDMS oil combined with fluorescence spectroscopy for the detection of UV fluorescent pollutants in aqueous solutions is a suitable analytical approach. The spectra collected by using self-designed equipment display a high similarity with the spectra obtained by the standard analytical procedure. The characteristic structure of the emission signal of phenanthrene was observed within the same area of the emission spectra and the comparison to the standard solution allowed its identification. The aim of these measurements was not the quantification but the qualitative detection of phenanthrene from contaminated samples. Furthermore it was revealed that EPDM rubber frames are not a suitable material for manufacturing PDMS-based devices, since it contains compounds which cause interference with the emission spectra of the target molecule. Under the tested conditions the saturation of the device was reached within 9 hours of exposure to contaminated solutions. Under conditions more close to practice (e.g. lower volume ratio between the PDMS and the water phase), the device might display much faster kinetics. Although the front face spectroscopy proposed here displays a high potential for development of this kind of monitor devices, is a more complicate procedure than the classical spectroscopy and it is not yet ready for on field applications. In conclusion, at the present time an ideal monitor device based on PDMS oil for the detection of PAHs in wastewater and surface water shall be designed for classical UV-fluorescence spectroscopy equipment. In this regard, the next part of the work was focused on the optimized design of a more suitable prototype PDMS device. The aim was to provide a more efficient working mechanism than the original device.

Two possible versions are proposed for this device. One for smaller sewers pipes 200/500 mm diameter (Figure 26) and one by-pass version for combined sewers catchments or industrial discharge pipes, with a bigger diameter (Figure 27). The model shown in Figure 25 was chosen for the elaboration of a predictive sorption curve, which is useful to estimate its accumulation potential for phenanthrene from water. Using the partition coefficients of phenanthrene between PDMS materials and water, available in literature, the absorption curve of phenanthrene onto the PDMS device during the exposure in contaminated aqueous systems was simulated and is presented in Figure 25. The volume of PDMS oil exposed to the contaminated flow of water is 0.96 dm^3 and all the calculations were carried out on the basis of this specific size. This device is shown in Figure 25 and would be used for on-line measurements.

3 Results

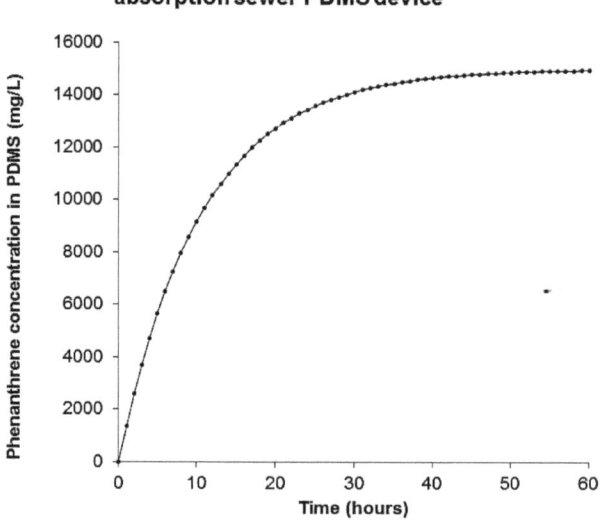

Figure 24 Aborption model for PDMS device. The absorption of phenanthrene into the PDMS device was estimated using the PDMS-water partition coefficient proposed by Sprunger et al., 2007 (Log $K_{PDMS-water}$ = 4, k = 10000).

This simulation curve is based on the partition coefficient of phenanthrene between water and PDMS and its diffusivity in PDMS calculated by Sprunger and co-workers, using the solvation parameter model of Abraham (1 x 10^{-6} cm^2/sec, Sprunger et al., 2007). The partition coefficient of phenanthrene between water and PDMS ($K_{PDMS-water}$) is equal to 10000. This means that the equilibrium concentration of phenanthrene inside a PDMS phase, after the partitioning process took place, is 10000 times higher than the concentration in water. Considering a constant concentration of 1.5 mg/L in the water phase, it was estimated that after 30 hours, close to the equilibrium, the concentration inside the chosen device is equal to 14100 mg/L. According to the physical-chemical properties of PDMS, no significant desorption of phenanthrene from the PDMS device was accounted between 20° C and 30° C for a period of several weeks. The desorption occurs but so slowly that is negligible for practical purposes. Faster desorption kinetics can be achieved increasing the temperature over 100 °C. This is a process expressed by the van't Hoff equation, in which the partition

3 Results

coefficient of a compound between two immiscible phases depends on the temperature of the system. In the case of phenanthrene and PDMS the value of the partition coefficient does not change significantly up to 80-100 ° C (DiFilippo and Egahouse 2010). In this case the desorption is not considered significant since the water temperature in combined sewers is usually about 20°C-30°C.

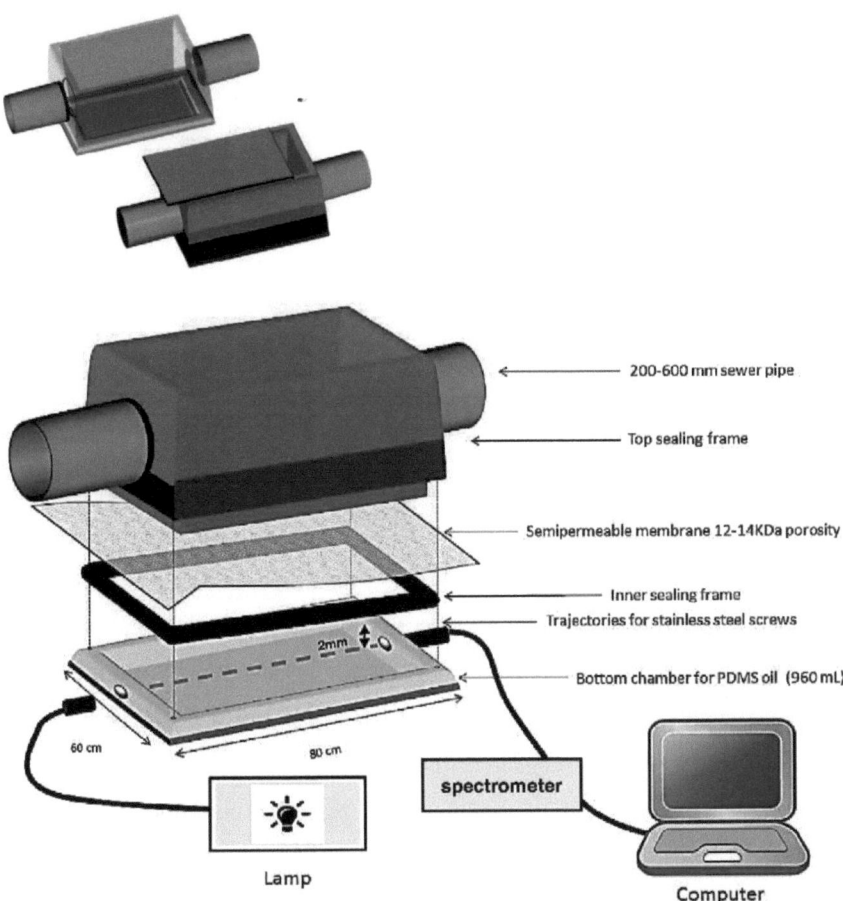

Figure 25 New prototype for monitoring hydrophobic pollutants in surface water and wastewater streams. The device is designed for small sewers with diameter 200 mm, which is the minimum size according to the BS EN758:2008 regulation. It is composed of two external frames (stainless steel) which enclose hermetically (by use of screws) a PDMS oil chamber placed on the bottom of the device. The dot lines in the picture show the trajectory of the screws which link together the external frame and the bottom chamber. In order to ensure the complete closure of the inner chamber, a tin frame will be placed between the bottom chamber and the semipermeable membrane. Once the device is completely assembled, it will be

3 Results

constantly filled with wastewater and the gel on the bottom of the device will extract the pollutants from the water. Using fiber optic devices the oil will be analyzed from the bottom chamber and the emission spectra will be collected by use of specific softwares. It will be analyzed by fluorescence spectroscopy using portable equipment already available on the market.

Figure 26 Combined sewer by-pass monitor device. The whole device is presented here completely assembled as a by-pass model prototype. The PDMS device presented before in Figure 25 is now linked to the wastewater stream by tubes. The use of pumps might be needed for keeping the low pressure inside the device and avoid damages.

According to this new design, a volume (3.5 mL) of the PDMS oil is analyzed from the bottom of the device by on-line connected UV fluorescence equipment. The UV light beam would pass across the all chamber and would be collected by fiber optic devices connected to a detector and a computer. This equipment is already available on the market and is used for similar on field measurements, therefore the development of such a technology would be easier to perform than the front-face application. The PDMS oil is a very interesting alternative to all the sorbent materials used for passive accumulation of pollutants in aqueous environments. It is a completely UV transparent material and therefore it allows on field measurements, which are faster than most of the laboratory procedures based on the extraction of the target compound from the sorbent phases.

4 DISCUSSION

4.1 Sewer biofilms as memory of PAHs in sewers

Wastewater, surface water and ground water are among the main carriers of pollutants in the open environment. Strongly hydrophobic pollutants are partitioned through the absorption into various organic phases such as soil, sediments, sludge, living organisms and microbial biofilms. Sewers are one of the main infrastructures for the distribution of wastewater. It has been reported that microbial biofilms grow abundantly on the inner surface of sewer pipes and in the sediments deposited on the bottom layer (Jahn and Nielsen 1998; Flemming and Wingender 2010). Microbial biofilms, as sorbent phases, are involved in the fate of pollutants in all aqueous systems (Spaeth et al., 1998, Headley et al., 1998), hence also in sewers (Rocher et al., 2003). In fact, various authors reported that microbial biofilms such as also sewer biofilms display high sorption capacity for hydrophobic organic compounds present in the water phase (Spaeth et al., 1998, Wicke et al., 2007, Antusch et al., 1995), but only few exploited biofilms for monitoring the pollution in sewers, following a systematic sampling strategy (Genuit 2008; Genuit and Block 2009). One of the aims of this study was to establish a monitoring procedure for sampling microbial biofilms from the sewers and analyze them for tracing the primary sources of polycyclic aromatic hydrocarbons. In order to achieve this target it was necessary to develop sampling protocols and to chose a suitable analytical method. In collaboration with the environmental department of the Bielefeld city hall, with the group of Dr. Gerard Genuit, the sampling of the biofilms was carried out using a plastic stripe support on which biofilm could grow, for a period of 4-6 weeks (Genuit 2008; Genuit and Block 2009). The analysis of the biofilm samples was carried out in order to detect phenanthrene, as marker compound for PAH pollution (Kuusimaki et al., 2004; Larsen and Baker 2003). The biofilms suspensions were analyzed after a liquid-liquid extraction step and by GC-MS analysis of the extracted solutions (materials and methods, page 68). This part of the work was carried out following the protocol DIN

4 Discussion

38414-23, which was chosen according to the available instrumentation and according to previous sudies (Lepom et al., 2009, Krüger et al., 2012, Bercaru et al., 2006). From the analyses, no significant trace of phenanthrene was observed in the extracts. Phenanthrene was expected to be detected in the biofilm samples, since they were exposed to sources such as kitchens, industrial activities and mixed urban discharges. As mentioned in the introduction, these activities can contribute significantly to the emission of PAHs in the environment.

With respect to these results, few explanations might be proposed. First of all, it is possible that no discharge of phenanthrene in the wastewater occurred during the monitoring period within the monitored area. In other words, no illicit emission of hazardous untreated effluents was carried out in the sampling area.

Another explaination may be that, although the discharges did occur, no significant retention of phenanthrene has been displayed from the biofilm layer grown over the plastic support. As previously reported, the partitioning of smaller PAH compounds in aqueous systems occurs mainly by sorption to particulate matter (PM, Kim and Kwon 2010), which is characteristic of combined wastewater streams. Particulate matter is a major constituent of the sediments accumulated on the bottom of the pipes. These latter ones could display higher organic carbon content, higher distribution coefficient and higher density than the biofilms developed on the support; these properties might turn them in preferred sinks for phenanthrene.

Only few studies have been focused on the desorption kinetics of organic pollutants in water biofilms under dynamic conditions. Although they might be not fully focused on wastewater streams, I will review some of those, which best suit the present essay.

In 1997 Schorer and colleagues stated that there was a lack of information related to the sorption/desorption of organic pollutants in river biofilms and that the knowledge about these processes would be crucial in understanding the fate of pollutants in water bodies. They monitored the concentration of polycyclic aromatic hydrocarbons (PAH) and polychlorinated biphenyls (PCBs) in biofilms of a contaminated German

4 Discussion

river close to Trier (Schorer et al., 1997). They noticed a pattern which consisted of repeating sorption peaks followed by immediate decrease of compound concentration in the biofilm sample. A period of less than 5 days was observed to be sufficient for significant changes in the concentration of Benzo(a)pyrene inside the biofilms. The authors explain that these decreases can be mainly due to the desorption of the pollutants form the biofilm. This is a prove of the reversible sorption mechanism of organics in biofilms and may explain what was observed in this study. After this, only few other studies were carried out on the same subject.

Headley and colleagues used a rotating annular reactor for studying the sorption of triallate, diclofop-methyl and atrazine to water biofilms (Headley et al., 1998). They observed fast absorption kinetics within the first 10 minutess of exposure to the contaminated solution and then slower desorption of the pollutants from the biofilm. The desorption equilibrium was reached after 150-200 minutes and involved the 50-90% of the absorbed pollutant (Headley et al 1998). During this experiment the Elbe river water was flushed into the reactor from a 40 L reservoir not in a circular flow and the effluent was collected outside from the reactor. In this study the flow was not circulating in a closed loop system and therefore the conditions were closer to practice if compared with most of the other studies where the desorption kinetics were measured under shaking batch conditions. However, the water concentration of the pollutants in the inlet flow was kept constant and it was not possible to observe the effect of short time exposure of the biofilm to a contaminated flow of aqueous solution. In fact, under real conditions the exposure of the biofilm to the contaminated flow would be limited to a short period of time and the exposure to a subsequent flowing clean water volume may cause faster desorption mechanisms than those observed by the authors of this study.

Lawrence and colleagues in 2001 reported the sorption of atrazine and diclofop onto biofilms grown in a rotating annular reactor flushed with the water of the South Saskatchewan River (Canada). During the sorption experiments the contaminated solution of atrazine or diclofop was let run at 12 L/h. The decrease of the chemicals from the bulk water was interrupted by pulsing increases of concentrations, which are assigned by the authors to the desorbed pollutant molecules from the biofilm in the reactor. This phenomenon is observed mainly between the second and the fifth day of exposure. The absorption equilibrium occurred after about 24 hours. If the hypothesis of the author is valid, the desorption begins right after that the absorption reached the equilibrium and it would display kinetics similar to the absorption process, desorbing 50% of the absorbed diclofop (Lawrence et al., 2001). However, also in this case the experimental system does not fully mimic real conditions. The experiments were carried out using a closed loop system and the water phase was changed with new contaminated water every 1, 3, 5, 8 days. Under real conditions, the water volume in contact with the biofilm surface would be replaced by clean water at higher rates. Therefore, it is possible that under real flowing conditions, the desorption of diclofop would be significantly increased, determining a rather short memory of a specific biofilm portion for this pollutant.

A short memory would not allow to detect phenanthrene in the biofilm samples using a 4 weeks sampling schedule, as I did during during my work. According to the reviewed material, the sorption of hydrophobic compounds in biofilms under dynamic conditions, it is a reversible mechanism. This might be the actual explaination of the results of this study and needs to be further investigated.

In conclusion, it must be pointed out that despite the results of this work and some other studies previously reported, other authours found sewer biofilms to be suitable for monitoring organic pollutants in sewers (Koch 2003; Antusch et al., 1995) and therefore a more complex explaination must be formulated. Microbial biofilms are highly heterogeneous matrices and no biofilm is equal to another. All results on this subject shall be considered partial.

4.2 PDMS device for detection of phenanthrene in sewers

Due to the results obtained by using sewer biofilms as passive samplers for phenanthrene, it became necessary to develop an alternative analytical method. A device was manufactured in order to achieve this goal. It was designed for two main

4 Discussion

purposes: passive accumulation of the target compound from the aqueous phase and detection of the pollutant directly from the device using optical fibers and UV fluorescence spectroscopy equipment. The device was submerged both in a deionized water solution of phenanthrene (1.3-1.5 mg/L) and in a natural pond water suspension spiked with phenanthrene (1.3-1.5 mg/L).

Passive accumulation devices (PAD) have been widely used in monitoring pollutants in aqueous systems and various examples have been reported in literature (Stuer-Lauridsen 2005). Some of them are shown in Table 19.

Table 19 Passive accumulation devices reviewed in Stuer-Lauridsen 2005. Various sorbent/membrane combinations have been tested in each work.

Sorbent material/membrane	Target compounds	Measuement
Trimethylpentane/polyethylene	Chlordane and dieldrin	n.a.
Polyurethane/fibreglass	Aromatic compounds	GC-MS
Low-density polyethylene	PAHs and PCBs	GC-mass selective detector
Dowex Optipore L-493/ceramic membrane	BTEX and naphthalenes	GC-mass selective detector
Hexane/polyethylenemembrane	Organochlorines	GC-MS
XAD-7 TenaxTa/ /polycarbonatemembrane	Phenolic compounds	n.a.

In the last 15 years many PADs have been constructed for the detection of hydrophobic pollutants in aqueous systems. Most of these devices are designed for 1-4 weeks field deployment, where uptake is governed by first order kinetics providing a time weighted average of the exposure concentration (Stuer-Lauridsen 2005). Semipermeable membrane devices (SPMD) are the most used amongst the PADs. Here follows brief explanation of their functioning. A chemical compound in the water is carried to the sampler by convection; it diffuses across the boundary phase surrounding the sampler, and passes through the membrane pores by conduction. It is finally solubilized in the solvent or sorbed to a bonded receiving phase. The final phase is chosen to act as a sink for the chemical, thus ensuring an effective gradient across the sampler's interface to the ambient water. Obviously, the effectiveness of the sampler is related to the surface area, and to increase this factor, several of the samplers allow the membrane to completely enclose the receiving phase forming a bag, tube or sandwich, such as a permeable housing (Stuer-Lauridsen 2005).

Different solvents and membranes can be used. Polyethylene, cellulose or ceramic membranes are the most common ones. Polyethylene is the preferred one for the accumulation of hydrophobic pollutants, since it displays more hydrophobic interactions and separates better the inner phase of the device from the water phase. Sorbent phases can greatly vary from solvents such as hexane, trimethylpentane, triolein to resins such as Dowex Optipore L-493 (Stuer-Lauridsen 2005).

Polydimethylsyloxane (PDMS) is a UV transparent material that is widely used for microfluidic chip analyses. Many PDMS devices that use fluorescence spectroscopy have been described (Wieder et al., 2005). PDMS can be found as additive in food, cosmetics, can be added as antifoam agent in lubricant solutions and in medicine for various purposes such as skin topical applications or wound dressings (ISO 10993, USP and European monographs). It is harmless for living organisms and the environment. From the wasterwater, PDMS accumulates in the sewage sludge. When this sludge is used for landfilling, PDMS undergoes processes of fast soil hydrolysis (Graiver, 2003). These degradative processes produce low molecular weight silanols, which are volatile. In the athmosphere, these compounds are degraded by OH radicals as shown in Graiver et al., 2003. In this study the aim was to create a prototype analytical system that could be used for on field measurements by front face UV-fluorescence spectroscopy. PDMS is highly hydrophobic and it is widely used as coating of solid phase micro extraction (SPME) fibers for detecting hydrophobic compounds in environmental samples (ter Laak et al., 2006; Yang et al., 2007). In this study PDMS oil was used for testing a prototype monitor device for PAH pollutants in sewers and surface water streams. The featuring of a passive accumulation device (PAD) containing PDMS oil with a front-face fluorescence spectroscopy method is a innovative idea for monitoring pollutants in water bodies.

4 Discussion

Wang et al., 2001 reported the use of a SPMD based on triolein and enclosed with polyethylene tubing membrane for monitoring aromatic and chlorinated compounds within the wastewater treatment plant of Beijing city. The use of this polymeric membranes is inspired on the interactions which occur at the level of biological membranes of living organisms exposed to the contaminated water stream, such as fish, mosses or algae. All the SPMD analysis, reported by Wang and co-workers, involves an extraction procedure to take out the absorbed pollutant and analyze it by GC-MS and other methods. The advantage of using PDMS resides in its UV transparency. Using a suitable excitation light beam, it might be possible to detect a target PAH compound absorbed by the PDMS oil contained into the device.

The preliminary results obtained testing this device in deionized water and pond water, show that the use of PDMS oil for front face fluorescence detection of phenanthrene in contaminated suspensions is a promising application. In the experiment carried out in the deionized water spiked with phenanthrene, mixing (900 rpm) and a volume ratio 1:25 between the device and the batch solution of phenanthrene were kept constant. The saturation of the device was reached after 9 hours and the signal of phenanthrene was well defined, considering this as a preliminary test. The optical fiber platform used for this experiment displayed good efficiency in collecting the emission signal from the device (pages 80-83). The heterogenous composition of the pond water suspension, during 24 hours of exposure, didn't influence the detection of phenanthrene into the device. Other hydrophobic molecules were expected to be partially absorbed into the PDMS, but the possible interferences didn't compromise the detection of the phenanthrene signal. After one hour of exposure it was already possible to observe the presence of phenanthrene inside the PDMS oil, by the arising of the first fluorescence peak of phenanthrene (345-350 nm). Unfortunately, the high fluorescence baseline didn't allow an optimal resolution of the signal. The explanation for such an effect might be assigned to a constituent of the sealing frames: ethylene propylene diene monomer (EPDM). The contact between the PDMS oil and this material, caused the chemical extraction of some hydrophobic substance from the frames and the arising of a

4 Discussion

fluorescence broad peak around 330 nm. This peak caused the high baseline and affected the estimation of the peak areas of the phenanthrene signal. For further optimization of the device, metal frames, tin or cupper preferably, shall be used in order to avoid this kind of inconvenience. The composition of the EPDM used is unknown and only some hypothesis can be formulated. EPDM is produced in a coordinative, anionic polymerization of ethylene, propylene and a non-conjugated diene such as hexadiene, dicyclopentadiene or ethylidene norbornene in a solution, suspension or gas phase technique (Röthemeyer and Sommer, 2006). Plasticizers are not usually present in EPDM rubber.

Only for specific purposes they are added in the EPDM mixture (Wypych 2004):

- Pressure sensitive tapes used to join rubber membranes.
- Hose formulation.
- Cold shrinkable cable joint protection.
- Weather-stripping composition.

The most common plasticizers used for EPDM are (Wypych 2004):

- Polyisobutylene (the most frequently used non-migrating plasticizer of EPDM compositions).
- Paraffin oil.
- Dibutyl phthalate.
- Dioctyl phthalate.
- Vulcanized vegetable oil.

Due to their fluorescence properties, these compounds were not considered the cause of the observed interferences on the emission spectrum of phenanthrene. Other compounds of the EPDM rubber materials were investigated. Since various components of EPDM display absorbance in the UV range and are photo oxidized by UV light, in order to prevent the aging of EPDM material, anti-oxidants are often added to the matrix (Rivaton et al., 2005). These compounds absorb UV light in the range of 220-280 nm, and therefore these compounds might have caused the interference and the broad emission signal from 320 nm to 360 nm.

The use of steady state fluorescence spectroscopy for the detection of a specific compound in complex solutions presents limitations when more than one compound can be excited at the same wavelength. Interferences are one of the main limits of fluorescence spectroscopy when it comes to multicomponent mixtures such as environmental samples. Although this limitation seems rather critical to overtake, some alternative approaches can be adopted.

Using steady state spectroscopy, the simplest approach is to investigate both the emission and the excitation spectra of a mixture. In this way, it is possible to distinguish different compounds, which display overlapping in the emission spectra at the same excitation wavelength. When this simple procedure is not sufficient, there are few other possible solutions in order to identify single compounds.

The excitation emission matrix fluorescence (EEMF) is a method based on the principle just mentioned. It allows plotting emission intensities at all combinations of excitation and emission wavelengths in a single three-dimensional graph (Patra 2003). The measurement is usually made by selecting one excitation wavelength and scanning the emission wavelength over the region of interest. Repeating this process for more than one excitation wavelengths, provides data that can be presented as a three-dimensional surface, where x-axis is the emission wavelength, y-axis is the excitation wavelength and z-axis is the fluorescence intensity (Patra 2003).

The use of pulsed laser technology can be exploited for time-resolved measurements in order to identify a specific compound in complex mixtures. The fluorescence life-time is the time needed for a fluorophore in the excited state to reach the ground energetical state. Each compound is characterized by a specific fluorescence life-time, therefore the comparison between the decay rates of the fluorescence signal can lead to the identification of a specific molecule in solutions composed by different fluorophores (Patra et al., 2003).

4 Discussion

Another possible approach is the synchronous fluorescence scan (SFS). In this case both the excitation and the emission monochromators are scanned simultaneously. When the excitation and the emission profiles of a compound are known, then they can be compared with the data obtained by this method and the identification can be accomplished (Patra and Mishra 2001). All these analytical methods can increase the efficiency of the fluorescence spectroscopy as reference approach for the detection of different PAHs accumulated in the PDMS device.

As previously said this prototype was not designed only for the accumulation of phenanthrene from the water phase. It was manufactured also to be suitable for front face spectroscopic measurements. In particular this function is based on the surface optical properties of the dialysis membrane and this is a complex parameter that needs to be further optimized in order to display a higher degree of resolution. This is the reason why a new prototype device, inspired by this prototype, (shown in Figure 26 page 90) was conceived and designed. This device represents a new idea for on field applications. Its high potential was presented here for the first time and might stimulate the interest of further investments. Once this new version of the PDMS device was designed, its absorption curve has been simulated under conditions close to practice, using the Fick's laws equations. As a result of this theoretical simulation, the exposure of 10 hours to a continuous flow of water saturated with phenanthrene (1.5 mg/L), would lead to an accumulation of about 8000-10000 mg/L of phenanthrene into the given PDMS oil phase. At a temperature comprised between 30° C and 60° C, the desorption of phenanthrene from the device into pure water would take so long (several weeks) that it is considered negligible whithin a period of several days. The desorption and absorption curves of phenanthrene into and from the device have been calculated considering deionized water as aqueous phase. It must be pointed out that, under real conditions, the presence in water of surfactants, particles and the deposition of sediments on the membrane of the device might significantly affect the sorption mechanism.

4 Discussion

The use of PDMS oil for the fluorescence detection of phenanthrene in sewers has been proposed here as an alternative application to sewer biofilms. The potential and the weaknesses of this application have been discussed, but thanks to this latter part of the work, it is now highlighted the path to a novel strategy for monitoring PAHs in sewers. PDMS oil seems to display the right properties for on field measurements.

In conclusion, this manuscript described theorethically and empirically the potential of sewer biofilms and PDMS oil for tracing hydrophobic pollutants in water systems.

In addition, the properties of polysaccharide gels have been introduced for inspiring new research on their use as sampler systems for tracking PAHs in sewers.

5 REFERENCES

Adav, S.S., Lee, D.J. (2008), Extraction of extracellular polymeric substances from aerobic granule with compact interior structure, Journal of Hazardous Materials, 154, 1120–1126.

Aksu, Z., (2005), Application of biosorption for the removal of organic pollutants: a review, Process Biochemistry, 40, 997–1026.

Ambati, J., Canakis, C. S., Miller, J. W., Gragoudas, E. S., Edwards, A., Weissgold, D. J., Adamis, A. P. (2000). Diffusion of high molecular weight compounds through sclera. Investigative ophthalmology & visual science, 41, 1181-1185.

Ambrosoli, R., Petruzzelli, L., Minati, J.L., Marsan, F.A., (2005), Anaerobic PAH degradation in soil by a mixed bacterial consortium under denitrifying conditions, Chemosphere, 60, 1231–1236.

Antusch, E., Sauer, J., Ripp, C., and Hahn, H.H. (1995), Organische Schadstoffe in der Sielhaut. Gas, Wasser, Abwasser, 75: 1010–1016.

Arfsten, D.P., Schaeffer, D.J., Mulveny, D.C., (1996), The effects of near ultraviolet radiation on the toxic effects of polycyclic aromatic hydrocarbons in animals and plants: A review. , Ecotoxicology and Environmental Safety, 33, 1-24.

Armisen R, Galatas F (1987) , Production, properties and uses of agar, In McHugh DJ (ed.), Production and Utilization of Products from Commercial Seaweeds, Fisheries Technical Paper 288, Rome: 1–57.

Association Franc-aise de Normalisation (AFNOR NF T90-376), 2000.Qualite´ de l'eau—De´termination de la toxicite´ chronique vis a` vis de Ceriodaphnia dubia en 7 jours—Essai d'inhibition de la croissance de la population.

Baek, S.O., Field, R.A., Goldstone, M.E., Kirk, P.W., Lester, J.N., Perry, R., (1991), A review of atmospheric polycyclic aromatic hydrocarbons: sources, fate and behavior, Water, Air, and Soil Pollution, 60, 79–300.

5 References

Bendt, B., Fetsch, B., Heier, C., Rondorf, A., (2007), Detektive im Abwasserkanal, Wasser-/Abwassertechnik, , 2-4.

Bercaru, O., Ulberth, F., Emons, H., & Vandecasteele, C. (2006). Accurate quantification of PAHs in water in the presence of dissolved humic acids using isotope dilution mass spectrometry. Analytical and bioanalytical chemistry, 384, 1207-1213.

Bergqvist, P. A., Jegorova, I., Kaunelienė, V., & Žaliauskienė, A. (2007). Dissolved organochlorine and PAH pollution profiles in Lithuanian and Swedish surface waters. Bulletin of environmental contamination and toxicology, 79, 147-152

Beuling, E. E., Van Dusschoten, D., Lens, P., Van Den Heuvel, J. C., Van As, H., & Ottengraf, S. P. P. (1998). Characterization of the diffusive properties of biofilms using pulsed field gradient-nuclear magnetic resonance. Biotechnology and bioengineering, 60, 283-291.

Beveridge, T.J., Hughes, M.N. and Lee, H., (1997), Metal microbe interactions: contemporary approaches, Advances in Microbial Physiology, 38, 177-24.

Beyenal, H., & Lewandowski, Z. (2000). Combined effect of substrate concentration and flow velocity on effective diffusivity in biofilms. Water research, 34, 528-538.

Bezalel L ., Hadar Y. and Cerniglia, C. E., (1997), Enzymatic Mechanisms Involved in Phenanthrene Degradation by the White Rot Fungus Pleurotus ostreatus, Applied and Environmental Microbiology, 63, 2495–2501.

Blanchard, M., Teil, M. J., Ollivon, D., Legenti, L., & Chevreuil, M. (2004). Polycyclic aromatic hydrocarbons and polychlorobiphenyls in wastewaters and sewage sludges from the Paris area (France). Environmental Research, 95, 184-197.

Boonchan, S., Britz, M.L., Stanley, G.A., 2000, Degradation and mineralization of high-molecular-weight polycyclic aromatic hydrocarbons by defined fungal-bacterial cocultures, Applied and Environmental Microbiology, 66, 1007-1019

5 References

Borde, X., Guieysse, B., Delgado, O., Munoz, R., Hatti-Kaul, R., Nugier-Chauvin, C., Patin, H., Mattiasson, B., (2003), Synergistic relationships in algal–bacterial microcosms for the treatment of aromatic pollutants, Bioresource Technology, 86, 293–300.

Bouzige, M., Pichon, V., Hennion, M., (1998), On-line coupling of immunosorbent and liquid chromatographic analysis for the selective extraction and determination of polycyclic aromatic hydrocarbons in water samples at the ng l−1 level, Journal of Chromatography A, 823, 197–210.

Brooke, D.N., Dobbs, A.J., Williams, N., (1986), Octanol/water partition coefficients (P): Measurement, estimation, and interpretation, particularly for chemicals with P > 105, Ecotoxicology and Environmental Safety, 11, 251–260.

Bruner, K. A., Fisher, S. W., & Landrum, P. F. (1994). The role of the zebra mussel, Dreissena polymorpha, in contaminant cycling: I. The effect of body size and lipid content on the bioconcentration of PCBs and PAHs. *Journal of Great Lakes Research*, *20*(4), 725-734.

Bruzzoniti, M. C., Sarzanini, C., & Mentasti, E. (2000). Preconcentration of contaminants in water analysis. Journal of Chromatography A, 902, 289-309.

Bumpus, J.A., (1989), Biodegradation of polycyclic aromatic hydrocarbons by Phanerochaete chrysosporium, Applied and Environmental Microbiology, 61, 2631–2635.

Camel, V., (2000), Microwave-assisted solvent extraction of environmental samples, Trends in analytical chemistry, 19, 229-248.

Campiglia, A. D., Hueber, D. M., Vo-dinh, T., (1996), Analysis of Polycyclic Aromatic Compounds in Soil Samples Using Laser-Induced Phosphorimetry, Polycyclic Aromatic Compounds, , Polycyclic Aromatic Compounds, 8, 117-128.

Celmera, D., Oleszkiewicza, J.A., Cicekb, N., (2008), Impact of shear force on the biofilm structure and performance of a membrane biofilm reactor for tertiary hydrogen-driven denitrification of municipal wastewater, Water Research, 42, 3057 – 3065.

Cerniglia C.E., (1992), Biodegradation of polycyclic aromatic hydrocarbons, Biodegradation, 3, 351–368.

Cerniglia, C. E., & Heitkamp, M. A. (1989). Microbial degradation of polycyclic aromatic hydrocarbons (PAH) in the aquatic environment. Metabolism of polycyclic aromatic hydrocarbons in the aquatic environment, 41-68.

Chang, B.V., Chang, S.W., Yuan, S.Y., (2003), Anaerobic degradation of polycyclic aromatic hydrocarbons in sludge, Advances in Environmental Research, 7, 623–628.

Chang, K.F., Fang, G.C., Chen, J.C., Wu, Y.S., (2006), Atmospheric polycyclic aromatic hydrocarbons (PAHs) in Asia: A review from 1999 to 2004, Environmental Pollution, 142, 388-396.

Charalabaki, M., Psillakis, E., Mantzavinos, D., Kalogerakis, N., (2005), Analysis of polycyclic aromatic hydrocarbons in wastewater treatment plant effluents using hollow fibre liquid-phase microextraction, Chemosphere, 60, 690–698.

Chen, C. E., Zhang, H., & Jones, K. C. (2012). A novel passive water sampler for in situ sampling of antibiotics. Journal of Environmental Monitoring, 14(6), 1523-1530.

Cheng, Y., Tsao, C. Y., Wu, H. C., Luo, X., Terrell, J. L., Betz, J., ... & Rubloff, G. W. (2012). Electroaddressing functionalized polysaccharides as model biofilms for interrogating cell signaling. Advanced Functional Materials, 22(3), 519-528.

Chiou, C.T., (1985), Partition Coefficients of Organic Compounds in Lipid-Water Systems and Correlations with Fish Bioconcentration Factors, Environmental Science & Technology, 19, 57-62.

Chiou, C.T., Porter, P.E. and Schmeddlng, D.W., (1983), Partition Equilibria of Nonionic Organic Compounds between Soil Organic Matter and Water, Environmental Science & Technology, 17, 227–231.

5 References

Chiou, C.T., Schmedding, D.W. and Manes, M., (1982), Partitioning of organic compounds in octanolwater systems, Environmental Science & Technology, 16, 4-9.

Clar, E., (1964), Polycyclic Hydrocarbons, New York: Academic Press . LCCN 63012392.

D'Adamo, R., Pelosi, S., Trotta, P., Sasone, G., (1997) , Bioaccumula-tion and biomagnification of polycyclic aromatic hydrocar-bons in aquatic organisms., Marine Chemistry, 56, 45–49.

Dai, J., Xu, M., Chen, J., Yang, X., & Ke, Z. (2007). PCDD/F, PAH and heavy metals in the sewage sludge from six wastewater treatment plants in Beijing, China. Chemosphere, 66, 353-361.

De Bruijn, J., Busser, F., Seinen, W., Hermens, J. (1989), Determination of octanol/water partition coefficients for hydrophobic organic chemicals with the "slow-stirring" method, Environmental Toxicology and Chemistry, 8, 499–512.

Dechadilok, P., & Deen, W. M. (2006). Hindrance factors for diffusion and convection in pores. Industrial & Engineering Chemistry Research, 45, 6953-6959.

Di Fabio, S., Lampisa, S., Zanettia, L., Cecchia, F., Fatonea, F., (2013), Role and characteristics of problematic biofilms within the removaland mobility of trace metals in a pilot-scale membrane bioreactor, Process Biochemistry, 48, 1757–1766.

DiFilippo, E. L., & Eganhouse, R. P. (2010). Assessment of PDMS-water partition coefficients: implications for passive environmental sampling of hydrophobic organic compounds. Environmental science & technology, 44(18), 6917-6925.

Directive 2455/2001/EC, European parliament, Establishing the list of priority substances in the field of water policy and amending Directive, Official Journal of the European Communities, L 331/1.

Du Laing, G., Rinklebe, J., Vandecasteele, B., Meers, E., & Tack, F. M. G. (2009). Trace metal behaviour in estuarine and riverine floodplain soils and sediments: a review. *Science of the Total Environment*, *407*(13), 3972-3985.

Dunne, M.Jr.W., (2002), Bacterial Adhesion: Seen Any Good Biofilms Lately?, Clinical Microbiology Reviews, 15, 155–166.

Dynes, J.J., Lawrence, J.R., Korber, D.R., Swerhone, G.D.W., Leppard, G.G., Hitchcock, A.P., (2006), Quantitative mapping of chlorhexidine in natural river biofilms, Science of the Total Environment, 369, 369–383.

Eastcott, L., Shui, W.Y. and Mackay, D., (1988), Environmen-tally relevant physiochemical properties of hydrocarbons: a review of data and development of simple correlations, Oil and Chemical Pollution, 4, 191–216.

Eisler, (1987), Polycyclic Aromatic Hydrocarbon Hazards to Fish, Wildlife and Invertebrates: A Synoptic Review, Contaminant Hazard Reviews, Report No. 11, Biological Report 85(1.11).

Environment Canada, 1993. Protocole-Test de fluctuation. Laboratoire C&P (CSL).

Eom, I.C., Rast, C., Veber, A.M., Vasseur, P., (2007), Ecotoxicity of a polycyclic aromatic hydrocarbon (PAH)-contaminated soil, Ecotoxicology and Environmental Safety, 67, 190–205.

Fact sheet "Phenanthrene" by US Environmental Protection Agency, web, (2014), http://www.epa.gov/osw/hazard/wastemin/minimize/factshts/phenanth.pdf, June 2014.

Fan, L.-S., Leyva-Ramos R., Wisecarver K. D. and Zehner B. J., (1990), Diffusion of phenol through a biofilm grown on activated carbon particles in a draft-tube three phase fluidized-bed bioreactor., Biotechnology and Bioengineering, 35, 279-286.

Flemming, H. C., & Wingender, J. (2010). The biofilm matrix. Nature Reviews Microbiology, 8, 623-633.

5 References

Flemming, H.-C., (1995), Sorption sites in biofilms, Water Science and Technology, 32, 27-33.

Flemming, H.-C., Leis, A. (2002), Sorpion properties of biofilms., Encyclopedia of Environmental Microbiology, 5, 2958-2967.

Flemming, H-C and Wingender, J., (2010), The biofilm matrix., Nature Reviews Microbiology, 8, 623-633.

Frølund, B., Griebe, T., Nielsen, P.H., (1995), Enzymatic activ-ity in the activated-sludge floc matrix., Applied Microbiology and Biotechnology, 43, 755-761.

Fytili, D., Zabaniotou, A., (2008), Utilization of sewage sludge in EU application of old and new methods—A review, Renewable and Sustainable Energy Reviews, 12, 116–140.

Galambos, P., & Forster, F. K. (1998, January). Micro-fluidic diffusion coefficient measurement. In Micro Total Analysis Systems' 98 (pp. 189-192). Springer Netherlands.

Garcia-Pichel, F., Johnson, S.L., Youngkin, D., Belnap, J., (2003), Small-Scale Vertical Distribution of Bacterial Biomass and Diversity in Biological Soil Crusts from Arid Lands in the Colorado Plateau, Microbial Ecology, 46, 312–321.

Garnya, K., Neub, T.R., Horna, H., (2009), Sloughing and limited substrate conditions trigger filamentous growth in heterotrophic biofilms—Measurements in flow-through tube reactor, Chemical Engineering Science, 64, 2723—2732.

Gasperi, J., Gromaire, M.C., Kafi, M., Moilleron, R., Chebbo, G., (2010), Contributions of wastewater, runoff and sewer deposit erosion to wet weather pollutant loads in combined sewer systems, Water research, 44, 5875 -5886.

Genuit, G. (2008), Korrespondenz Abwasser, 55, 777-781.

5 References

Genuit, G., Block, M., 2009, Ermittung von Einlatern PFT-haltigen Abwassers durch Untersuchung der Sielhaut, Gewasserschutz-wasser-Abwasser, Aachen 2009, ISBN 978-3-938996-23-2.

Gibizova, V. V., Komarova, A. V., Sergeeva, I. A., Fedorova, K. V., & Petrova, G. P. Interactions Between Biomarkers and Main Blood Proteins.

Gigliotti, C., Brunciak, P.A., Dachs, J., Glenniv, T.R., Nelson, E.D., Totten, L.A., Eisenreich, S.J., (2002), Air-water exchange of polycyclic aromatic hydrocarbons in the New York, New Jersey, USA, Harbor Estuary, Environmental Toxicology and Chemistry, 21, 235–244.

Golmohamadi, M. (2013). Quantifying diffusion in biofilms: from model hydrogels to living biofilms.

Gotovac, S., Yang, C. M., Hattori, Y., Takahashi, K., Kanoh, H., & Kaneko, K. (2007). Adsorption of polyaromatic hydrocarbons on single wall carbon nanotubes of different functionalities and diameters. Journal of colloid and interface science, 314, 18-24.

Graiver, D., Farminer, K. W., & Narayan, R. (2003). A review of the fate and effects of silicones in the environment. Journal of Polymers and the Environment, 11, 129-136.

Guo, G., Wu, F., He, H., Zhang, R., Feng, C., Li, H., & Chang, M. (2012). Characterizing ecological risk for polycyclic aromatic hydrocarbons in water from Lake Taihu, China. Environmental monitoring and assessment, 184, 6815-6825.

Gutekunst, B., (1989) Praktische Erfahrungen und Ergebnisse aus Sielhautuntersuchungen zur Ermittlung schwermetallhaltiger Einleitungen, Korr. Abwasser, 36, 1367-1375.

Haritash, A.K., Kaushik, C.P., (2009) Biodegradation aspects of Polycyclic Aromatic Hydrocarbons (PAHs): A review, Journal of Hazardous Materials, 169, 1–15.

5 References

Harrison, E.Z., Oakes, S.R., Hysell, M., Hay, A., (2006), Organic chemicals in sewage sludges, Science of the Total Environment, 367, 481–497.

He, X., Francis, L., Leming, M. L., Dean, L. O., Lappi, S. E., & Ducoste, J. J. (2013). Mechanisms of Fat, Oil and Grease (FOG) deposit formation in sewer lines. Water research, 47(13), 4451-4459.

Headley, J. V., Gandrass, J., Kuballa, J., Peru, K. M., & Gong, Y. (1998). Rates of sorption and partitioning of contaminants in river biofilm. Environmental science & technology, 32(24), 3968-3973.

Holden, P. A., Hunt, J. R., & Firestone, M. K. (1997) Toluene diffusion and reaction in unsaturated Pseudomonas putida biofilms. Biotechnology and bioengineering, 56, 656.

Houhou, J., Lartiges, B.S., Montarges-Pelletier, E., Sieliechi, J., Ghanbaja, J., Kohler, A., (2009) Sources, nature, and fate of heavy metal-bearing particles in the sewer system, Science of the Total Environment, 407, 6052–6062.

Hsiau, P., Lo, S., (1998) Extractabilities of heavy metals in chemically-fixed sewage sludges, Journal of Hazardous Materials, 58, 73–82.

Hussar, E., Richards, S., Lin, Z.-Q., Dixon, R.P., Johnson, K.A., (2012) Human Health Risk Assessment of 16 Priority Polycyclic Aromatic Hydrocarbons in Soils of Chattanooga, Tennessee, USA, Water Air Soil Pollut, 223, 5535–5548.

International Standard Organization (ISO 11267), 1999. Soil quality inhibition of reproduction of Collembola (Folsomia candida) by soil pollutants.

International Standard Organization (ISO 11268-1), 1993. Soil quality effects of pollutants on earthworms (Eisenia fetida)—Part 1: determination of acute toxicity using artificial soil substrates.

International Standard Organization (ISO 11268-2), 1998. Soil quality effects of pollutants on earthworms (Eisenia fetida)—part 2: determination of effects on reproduction.

5 References

International Standard Organization (ISO 11269-2), 1995. Soil quality determination of the effects of pollutants on soil flora—Part 2: effects of chemicals on the emergence and growth of higher plant.

International Standard Organization (ISO 6341), 1996. Water quality determination of the inhibition of the mobility of Daphnia magna Straus (Ciadocera, Crustacea)—acute toxicity test.

International Standard Organization (ISO 8692), 1996. Water quality freshwater algal growth inhibition test with Scenedesmus subspicatus and Selenastrum capricornutum.

International Standard Organization (ISO NF EN 11348-3), 1999.Water quality determination of the inhibitory effect of water samples on the light emission of Vibrio fischeri (Luminescent bacteria test, part 3).

International Standard Organization (ISO/DIS 13829), 2000. Water quality determination of the genotoxicity of water and waste water using the umu-test.

Jahn, A., & Nielsen, P. H. (1998) Cell biomass and exopolymer composition in sewer biofilms. Water Science and Technology, 37, 17-24.

Jennings, A.A., (2012) Worldwide regulatory guidance values for surface soil exposure to carcinogenic or mutagenic polycyclic aromatic hydrocarbons, Journal of Environmental Management, 110, 82-102.

Jerina, D. M., Selander, H., Yagi, H., Wells, M.C., Davey, J. F., Mahadevan, V., Gibson, D. T., (1976) Dihydrodiols from anthracene and phenanthrene, Journal of the American Chemical Society, 98, 5988–5996.

Johnsen, A.R., Wick, L.Y., Harms, H., (2005) Principles of microbial PAH-degradation in soil, , Environmental Pollution, 133, 71–84.

Johnson, E. M., Berk, D. A., Jain, R. K., & Deen, W. M. (1996) Hindered diffusion in agarose gels: test of effective medium model. Biophysical journal, 70, 1017-1023.

Jorand, F., Bou-Bigne, F., Block, J.C. and Urbain, V., (1998) Hydrophobic/hydrophilic properties of activated sludge exopolymeric substances, Water Science and Technology, 37, 307-315.

Jouenne, T., Tresse, O., & Junter, G. A. (1994) Agar-entrapped bacteria as an in vitro model of biofilms and their susceptibility to antibiotics. FEMS microbiology letters, 119, 237-242.

Kenaga, E. E. and Goring, C. A. I., (1978) Relationship between Water Solubility, Soil Sorption, Octanol-Water Partitioning and Bioconcentration of Chemicals in Biota, Aquatic Toxicology Symposium, Proceedings of the American Society for Testing and Material, No. STP 707, 78-115.

Keyte, I.J., Harrisonz, R.M. and Lammel, G., (2013) Chemical reactivity and long-range transport potential of polycyclic aromatic hydrocarbons – a review, Chemical Society Reviews, 42, 9333-9391.

Kim, I. S., Stabnikova, E. V., Ivanov, V. N., (2000) Hydrophobic interactions within biofilms of nitrifying and denitrifying bacteria in biofilters, Bioprocess Engineering, 22, 285-290.

Kima, K.-H., Jahan, S.A., Kabir, E., Brown, R.J.C., (2013) A review of airborne polycyclic aromatic hydrocarbons (PAHs) and their human health effects, Environment International, 60, 71–80.

Koch, M. (2003). Quellenermittlung von Schadstoffen in kommunalen Abwässern und Sedimenten.

Kochany, J., Maguire, R. J., (1992) Abiotic transformations of polynuclear aromatic hydrocarbons and polynuclear aromatic nitrogen heterocycles in aquatic environments, Science Total Environment, 144, 17-31.

Kostel, J.A. ; Wang, H. ; St. Amand, A.L.; Gray, K.A., (1999) Use of a novel laboratory stream system to study the ecological impact of PCB exposure in a periphytic biolayer, Water research.

Krüger, O., Kalbe, U., Berger, W., Simon, F. G., & Meza, S. L. (2012). Leaching experiments on the release of heavy metals and PAH from soil and waste materials. Journal of hazardous materials, 207, 51-55.

Kukharchyk, T. I., Khomich, V. S., Kakareka, S. V., Kurman, P. V., & Kozyrenko, M. I. (2013). Contamination of soils in the urbanized areas of Belarus with polycyclic aromatic hydrocarbons. Eurasian Soil Science, 46, 145-152.

Kuusimäki, L., Peltonen, Y., Mutanen, P., Peltonen, K., & Savela, K. (2004) Urinary hydroxy-metabolites of naphthalene, phenanthrene and pyrene as markers of exposure to diesel exhaust. International archives of occupational and environmental health, 77, 23-30.

Larsen, R. K., & Baker, J. E. (2003) Source apportionment of polycyclic aromatic hydrocarbons in the urban atmosphere: a comparison of three methods. Environmental Science & Technology, 37, 1873-1881.

Lawrence, J. R., Kopf, G., Headley, J. V., & Neu, T. R. (2001). Sorption and metabolism of selected herbicides in river biofilm communities. Canadian journal of microbiology, 47(7), 634-641.

Lawrence, J. R., Wolfaardt, G. M., & Korber, D. R. (1994) Determination of diffusion coefficients in biofilms by confocal laser microscopy. Applied and environmental microbiology, 60, 1166-1173.

Lazarova, V. and Manem, J., (1995) Biofilm characterization and activity analysis in water and wastewater treatment, Water research, 29, 2227-2245.

Lear, G., Lewis, G.D., (2012) Microbial Biofilms: Current Research and Applications, Caister Academic Press, Norfolk, UK.

Lee, M. L.; Novotny, M.; Bartle, K. D., (1981) Analytical Chemistry of Polycyclic Aromatic Compounds, Academic Press: New York.

Lepom, P., Brown, B., Hanke, G., Loos, R., Quevauviller, P., & Wollgast, J. (2009). Needs for reliable analytical methods for monitoring chemical pollutants in surface water under the European Water Framework Directive. Journal of Chromatography A, 1216, 302-315.

Lerche, D., Sørensen, P.B., Sørensen Larsen, H., Carlsen, L., Nielsen, O.J., (2002), Comparison of the combined monitoring-based and modelling-based priority setting scheme with partial order theory and random linear extensions for ranking of chemical substances, Chemosphere, 49, 637–649.

Liu, H., & Fang, H. H. (2002). Characterization of electrostatic binding sites of extracellular polymers by linear programming analysis of titration data. *Biotechnology and bioengineering, 80*(7), 806-811.

Liu, J., Liu, G., Zhang, J., Yin, H., & Wang, R. (2012). Occurrence and risk assessment of polycyclic aromatic hydrocarbons in soil from the Tiefa coal mine district, Liaoning, China. Journal of Environmental Monitoring, 14, 2634-2642.

Ma, Y. G., Lei, Y. D., Xiao, H., Wania, F., & Wang, W. H. (2009) Critical review and recommended values for the physical-chemical property data of 15 polycyclic aromatic hydrocarbons at 25 C. Journal of Chemical & Engineering Data, 55, 819-825.

Mahfoud, C.A., El Samrani, A.G., Mouawad, R., Hleihel, W., El Khatib, R., Lartiges, B.S., Ouaini, N., (2009) Disruption of biofilms from sewage pipes under physical and chemical conditioning, Journal of Environmental Sciences, 21, 120–126.

Manodori, L., Gambaro, A., Piazza, R., Ferrari, S., Stortini, A. M., Moret, I., & Capodaglio, G. (2006). PCBs and PAHs in sea-surface microlayer and sub-surface water samples of the Venice Lagoon (Italy). Marine Pollution Bulletin, 52, 184-192.

Manoli, E., Samara, C., (1999), Occurrence and mass balance of polycyclic aromatic hydrocarbons in the Thesssalonoki sewage treatment plant, Journal of Environmental Quality, 28, 176–187.

5 References

Mansuy-Huault, L., Regier, A., & Faure, P. (2009). Analyzing hydrocarbons in sewer to help in PAH source apportionment in sewage sludges. Chemosphere, 75, 995-1002.

Martinsen, A., Storrø, I., & Skjårk-Bræk, G. (1992). Alginate as immobilization material: III. Diffusional properties. Biotechnology and bioengineering, 39, 186-194.

Mayer, M.C., Moritz, R., Kirschner, C., Borchard, W., Maibaum, R., Wingender, J., Flemming, H.-C., (1999) The role of intermolecular interactions: studies on model systems for bacterial biofilms, International Journal of Biological Macromolecules, 26, 3–16.

McConkey, B.J., Duxbury, C.L., Dixon, D.G., Greenberg, B.M., (1997) Toxicity of a PAH photooxidation product to the bacteria Photobacterium phosphoreum and the duckweed Lemna gibba: effects of phenanthrene and its primary photoproduct, phenanthrenequinone, Environmental Toxicology and Chemistry, 16, 892–899.

Meidinger, R.F., St. Germain, R.W., Dohotariu, V. and Gillispie, G.D., (1993) Fluorescence of aromatic hydrocarbons in aqueous solutions, Proceedings of the U.S. EPA/Air and Waste Management Association International Symposium on Field Screening Methods for Hazardous Wastes and Toxic Chemicals, Las Vegas, NV, 395-403.

Microbics, 1993. Mutatox Manual. AZUR Environmental, Carlsbad, CA.

Miller, M. J., & Allen, D. G. (2004) Transport of hydrophobic pollutants through biofilms in biofilters. Chemical engineering science, 59, 3515-3525.

Miller, M.M., Wasik, S.P., (1985) Relationships between octanol–water partition coefficient and aqueous solubility, Environmental Science & Technology, 19, 522–529.

5 References

Moret, S., Conte, L.S., (2000) Polycyclic aromatic hydrocarbons in edible fats and oils: occurrence and analytical methods., Journal of Chromatography A, 882, 245–253.

Morgenroth, E., Wilderer, P.A., (2000) Influence of detachment mechanisms on competition in biofilms, Water research, 34, 417-426.

Motelay-Massei, A., Ollivon, D., Garban, B., Chevreuil, M., Polycyclic aromatic hydrocarbons in bulk deposition at a suburban site: assessment by principal component analysis of the influence of meteorological parameters Atmos Environ, 37 (2003), pp. 3135–3146.

Motelay-Massei, A., Ollivon, D., Garban, B., Teil, M. J., Blanchard, M., & Chevreuil, M. (2004). Distribution and spatial trends of PAHs and PCBs in soils in the Seine River basin, France. Chemosphere, 55, 555-565.

Muñoz, R., Guieysse, B., & Mattiasson, B. (2003). Phenanthrene biodegradation by an algal-bacterial consortium in two-phase partitioning bioreactors. Applied microbiology and biotechnology, 61(3), 261-267.

Nagy, A. S., Simon, G., Szabó, J., & Vass, I. (2013). Polycyclic aromatic hydrocarbons in surface water and bed sediments of the Hungarian upper section of the Danube River. Environmental monitoring and assessment, 185, 4619-4631.

Naik, M.M., Pandey, A., Dubey, S.K., (2012) Biological characterization of lead-enhanced exopolysaccharide produced by a lead resistant Enterobacter cloacae strain P2B, Biodegradation, 23, 775-783.

Nielsen, P.H., Raunkjær, K., Henrik Norsker, N., Aagaard Jensen, N. and Hvitved-Jacobsen, T., (1992) Transformation of Wastewater in Sewer Systems — A Review, Water Science & Technology, 25, 17–31.

NIST, Phenanthrene, The National Institute of Standards and Technology (NIST) is an agency of the U.S. Department of Commerce (2011), http://webbook.nist.gov, 31-11-2011

5 References

Northcott, G.L., Jones, K.C., (2000), Experimental approaches and analytical techniques for determining organic compound bound residues in soil and sediment, Environmental Pollution, 108, 19-43.

Nousiainen, U., Torronen, R. and Hanninen, O., (1984), Differential induction of various carboxylesterases by certain polycyclic aromatic hydrocarbons in the rat, Toxicology, 32, 243-251.

Okedeyi, O. O., Nindi, M. M., Dube, S., & Awofolu, O. R. (2013). Distribution and potential sources of polycyclic aromatic hydrocarbons in soils around coal-fired power plants in South Africa. Environmental monitoring and assessment, 185, 2073-2082.

Oleszczuk, P. (2009). Application of three methods used for the evaluation of polycyclic aromatic hydrocarbons (PAHs) bioaccessibility for sewage sludge composting. Bioresource technology, 100(1), 413-420.

Oleszczuk, P., (2008) Sorption of phenanthrene by sewage sludge during composting in relation to potentially bioavailable contaminant content, Journal of Hazardous Materials, 161, 1330–1337.

Olivella, M. A., (2006) Polycyclic aromatic hydrocarbons in rainwater and surface waters of Lake Maggiore, a subalpine lake in Northern Italy, Chemosphere, 63, 116–131.

Orecchio, S. (2010). Assessment of polycyclic aromatic hydrocarbons (PAHs) in soil of a Natural Reserve (Isola delle Femmine) (Italy) located in front of a plant for the production of cement. Journal of hazardous materials, 173, 358-368.

Patra, D., (2003) Applications and New Developments in Fluorescence Spectroscopic Techniques for the Analysis of Polycyclic Aromatic Hydrocarbons, Applied Spectroscopy Reviews, 38, 155–185.

Patra, D., Mishra, A. K. (2001) Investigation on simultaneous analysis of multicomponent polycyclic aromatic hydrocarbon mixtures in water samples: a simple synchronous fluorimetric method. Talanta, 55, 143-153.

Pérez, S., la Farré, M., García, M.J., Barceló, D., (2001) Occurrence of polycyclic aromatic hydrocarbons in sewage sludge and their contribution to its toxicity in the ToxAlert® 100 bioassay, Chemosphere, 45, 705–712.

Plachá, D., Raclavská, H., Matýsek, D. and Rümmeli, M.H., (2009) The polycyclic aromatic hydrocarbon concentrations in soils in the Region of Valasske Mezirici, the Czech Republic, Geochemical Transactions, 10, 12–33.

Priester, J.H., Olson, S.G., Webb, S.M., Neu, M.P., Hersman, L.E. and Holden, P.A., (2006) Enhanced Exopolymer Production and Chromium Stabilization in Pseudomonas putida Unsaturated Biofilms, Applied and Environmental Microbiology, 72, 1988–1996.

Qian, Y., Posch, T., & Schmidt, T. C. (2011). Sorption of polycyclic aromatic hydrocarbons (PAHs) on glass surfaces. Chemosphere, 82, 859-865.

Ramesh, A., Walker, S.A., Hood, D.B., Guillén, M.D., Schneider, K. and Weyand, E.H., (2004) Bioavailability and Risk Assessment of Orally Ingested Polycyclic Aromatic Hydrocarbons, International Journal of Toxicology, 23, 301-333.

Ravindraa, K., Sokhia, R., Van Grieken, R., (2008) Atmospheric polycyclic aromatic hydrocarbons: Source attribution, emission factors and regulation, Atmospheric Environment, 42, 1494–1501.

Readman, J. W., Mantoura, R. F. C. and Rhead, M. M., (1984), The Physico-Chemical Speciation of Polycyclic Aromatic Hydrocarbons (PAH) in Aquatic Systems, Fresenius' Zeitschrift für Analytische Chemie, 319, 126-131.

Reid, D. (1991) Paris Sewers and Sewermen, BT Global London.

Reusch, W., 2013, https://www2.chemistry.msu.edu/faculty/reusch/virttxtjml/photchem.htm, 05.05.2013.

Rice, S. A., Koh, K. S., Queck, S. Y., Labbate, M., Lam, K. W., & Kjelleberg, S. (2005). Biofilm formation and sloughing in Serratia marcescens are controlled by quorum sensing and nutrient cues. *Journal of bacteriology*, *187*(10), 3477-3485.

Rivaton, A., Cambon, S., & Gardette, J. L. (2006) Radiochemical ageing of ethylene–propylene–diene elastomers. 4. Evaluation of some anti-oxidants. Polymer degradation and stability, 91, 136-143.

Robards, K., Haddad, P.R. and Jackson, P.E., (1994) Principles and practice of modern chromatographic methods, Academic Press London.

Rocher, V., Azimi, S., Moilleron, R., & Chebbo, G. (2003) Biofilm in combined sewers: wet weather pollution source and/or dry weather pollution indicator? Water Science & Technology, 47, 35-43.

Rogers, H.R., (1996) Sources, behaviour and fate of organic contaminants during sewage treatment and in sewage sludges, The Science of the Total Environment, 185, 3-26.

Röthemeyer F, Sommer F (2001) Kautschuktechnologie. München: C. Hanser

Samanta, S. K., Chakraborti, A. K., Jain, R. K., (1999) Degradation of phenanthrene by different bacteria: evidence for novel transformation sequences involving the formation of 1-naphthol, Applied Microbiology and Biotechnology, 53, 98-107.

Satoh, H., Odagiri, M., Ito, T., Okabe, S., (2009) Microbial community structures and in situ sulfate-reducing and sulfur-oxidizing activities in biofilms developed on mortar specimens in a corroded sewer system, Water research, 43, 4729 – 4739.

Schmitt, J., Nivens, D., White, D.C. and Flemming, H.-C., (1995) Changes of biofilm properties in response to sorbed substances - an FTIR-ATR study., Water Science and Technology, 32, 149-155.

Schorer, M., & Eisele, M. (1997). Accumulation of inorganic and organic pollutants by biofilms in the aquatic environment. Water, Air, and Soil Pollution, 99(1-4), 651-659.

5 References

Schüth, (1994) Sorptionskinetik und Transportverhalten von Polyzyklischen Aromatischen Kohlenwasserstoffen (PAK) im Grundwasser, DissertationUniversity of Tübingen, FRG (1994).

Semple, K. T., Morriss, A. W. J., Paton, G. I., (2003) Bioavailability of hydrophobic organic contaminants in soils: fundamental concepts and techniques for analysis, European Journal of Soil Science, 54, 809–818.

Seo, J.-S., Keum, Y.-S. and Li, Q.X., (2009) Bacterial Degradation of Aromatic Compounds, International Journal of Environmental Research and Public Health — Open Access Journal, 6, 278- 309.

Sharma, M., & Yashonath, S. (2007) Size Dependence of Solute Diffusivity and Stokes-Einstein Relationship: Effect of van der Waals Interaction. Diffusion Fundamentals, 7, 11-1.

Sheng, G.-P., Yu, H.-Q., Li, X.-Y., (2010) Extracellular polymeric substances (EPS) of microbial aggregates in biological wastewater treatment systems: A review, Biotechnology Advances, 8, 882–894.

Shuttleworth, K.L., Cerniglia, C.E., (1996) Bacterial Degradation of Low Concentrations of Phenanthrene and Inhibition by Naphthalene, Microbial Ecology, 31, 305-317.

Simmon, V.F., Rosenkranz, H.S., Zeiger, E. and Poirier, L.A., (1979) Mutagenic Activity of Chemical Carcinogens and Related Compounds in the Intraperitoneal Host-Mediated Assay, Journal of the national cancer institute, 62, 911-918.

Singh, A., Ward, O.P., (2004) Biodegradation and Bioremediation, Springer-Verlag GmbH, Berlin.

Singh, R., Paul, D. and Jain, R.K., (2006) Biofilms: implications in bioremediation, Trends in Microbiology, 14, 389-397.

Singh, S.N., (2012) Microbial Degradation of Xenobiotics, Environmental Science and Engineering, Springer Heidelberg Dordrecht London New York.

Song, Y.F., Jing, X., Fleischmann, S., Wilke, B.-M., (2002) Comparative study of extraction methods for the determination of PAHs from contaminated soils and sediments, Chemosphere, 48, 993–1001.

Spath, R., Flemming, H.-C. and Wuertz, S., (1998) Sorption properties of biofilms, Water Science and Technology, 37, 207-210.

Sprunger, L., Proctor, A., Acree Jr, W. E., & Abraham, M. H. (2007). Characterization of the sorption of gaseous and organic solutes onto polydimethyl siloxane solid-phase microextraction surfaces using the Abraham model. Journal of Chromatography A, 1175, 162-173.

Srogi, K., (2007) Monitoring of environmental exposure to polycyclic aromatic hydrocarbons: a review, Environmental Chemistry Letters, 5, 169–195.

Strathmann, M., Leon-Morales, C.F., Flemming, H.-C., (2007) Influence of Biofilms on Colloid Mobility in the Subsurface, Colloidal Transport in Porous Media, 2007, 143-173.

Stuer-Lauridsen, F. (2005) Review of passive accumulation devices for monitoring organic micropollutants in the aquatic environment. Environmental Pollution, 136, 503-524.

Sutherland, I. W. (2001) Biofilm exopolysaccharides: a strong and sticky framework. Microbiology, 147, 3-9.

Sverdrup, L.E., Nielsen, T., Krogh, P.H., (2002) Soil Ecotoxicity of Polycyclic Aromatic Hydrocarbons in Relation to Soil Sorption, Lipophilicity, and Water Solubility, Environmental Science & Technology, 36, 2429-2435.

ter Laak, T. L., Agbo, S. O., Barendregt, A., & Hermens, J. L. (2006) Freely dissolved concentrations of PAHs in soil pore water: Measurements via solid-phase extraction and consequences for soil tests. Environmental science & technology, 40, 1307-1313.

5 References

The Sewage Sludge Directive 86/278/EEC, council directive on the protection of the environment, and in particular of the soil, when sewage sludge is used in agriculture, Official Journal L 181, 04/07/1986 P. 0006 – 0012.

Tresse, O., Jouenne, T., & Junter, G. A. (1995) The role of oxygen limitation in the resistance of agar-entrapped, sessile-like Escherichia coli to aminoglycoside and β-lactam antibiotics. Journal of Antimicrobial Chemotherapy, 36, 521-526.

Trzesicka-Mlynarz, D., & Ward, O. P. (1995). Degradation of polycyclic aromatic hydrocarbons (PAHs) by a mixed culture and its component pure cultures, obtained from PAH-contaminated soil. Canadian journal of microbiology, 41, 470-476

Tsezos, M. and Bell, J. P., (1988) Comparison of the biosorption and desorption of hazardous organic pollutants by live and dead biomass, Water research, 23, 561-568.

US EPA, (1993), Standards for the Use or Disposal of Sewage Sludge; Final Rules, Federal Register Notice (FRN), 58 FR 9248, 40 CFR Parts 257, 403, 503, Friday, February 19, 1993.

US EPA, (2001), Integrated Risk Information System. Online. Office of Health and Environmental Assessment, National Center for Environmental Assessment, Cincinnati, OH.

USGS, (2009), Octanol-Water Partition Coefficient (KOW), Definitions, http://toxics.usgs.gov/definitions/kow.html, 02-Jun-2014 17:59:48 EDT.

Van der Bruggen, B., Schaep, J., Wilms, D., & Vandecasteele, C. (1999) Influence of molecular size, polarity and charge on the retention of organic molecules by nanofiltration. Journal of Membrane Science, 156, 29-41.

van Hullebusch, E.D., Zandvoort, M.H., Lens, P.N.L., (2003) Metal immobilisation by biofilms: Mechanisms and analytical tools, Reviews in Environmental Science and Bio/Technology, 2, 9–33.

Vigneswaran, S., Davis, C., Kandasamy, J. and Chanan, A., (2009) Urban Wastewater Treatment: Past, Present and Future, Water and Wastewater Treatment Technologies, Vol.1.

Vincke, E., Boon, N., Verstraete, W., (2001) Analysis of the microbial communities on corroded concrete sewer pipesa case study, Applied Microbiology and Biotechnology, 57, 776–785.

Vogt, M., Flemming, H. C., & Veeman, W. S. (2000) Diffusion in< i> Pseudomonas aeruginosa</i> biofilms: a pulsed field gradient NMR study. Journal of biotechnology, 77, 137-146.

Wang, K.; Hu, Y.; Liu, Y.; Mi, N.; Fan, Z.; Liu, Y.; Wang, Q., (2010) Design, synthesis, and antiviral evaluation of Phenanthrene-based tylophorine derivatives as potential antiviral agents., Journal of Agriculture and Food Chemistry, 58, 12337-12342.

Wang, W., Wang, W., Zhang, X., & Wang, D. (2002). Adsorption of p-chlorophenol by biofilm components. *Water Research*, *36*(3), 551-560.

Wang, Z., Friedrich, D. M., Beversluis, M. R., Hemmer, S. L., Joly, A. G., Huesemann, M. H., Peyton, B. M. (2001) A fluorescence spectroscopic study of phenanthrene sorption on porous silica. Environmental science & technology, 35, 2710-2716.

Wanga, W., Wangb, W., Zhanga, X., Wang, D., (2002) Adsorption of p-chlorophenol by biofilm components, Water Research, 36, 551–560.

Water Frame Directive (WFD) 2000/60/EC, The EU Water Framework Directive - integrated river basin management for Europe.

Wei, L., Brossi, A., Kendall, R., Bastow, K.F., Morris-Natschke, S.L., Shi, Q., Lee, K.H., (2006) Antitumor agents 251: Synthesis, cytotoxic evaluation, and structure–activity relationship studies of phenanthrene-based tylophorine derivatives (PBTs) as a new class of antitumor agents, Bioorganic & Medicinal Chemistry, 14, 6560–6569.

5 References

Westrin, B. A., & Axelsson, A. (1991) Diffusion in gels containing immobilized cells: a critical review. Biotechnology and bioengineering, 38, 439-446.

White, C., Gadd, G.M., (1998), Accumulation and effects of cadmium on sulphate-reducing bacterial biof ilrns, Microbiology, 144, 1407-1415.

Wicke, D., Böckelmann, U., & Reemtsma, T. (2007) Experimental and modeling approach to study sorption of dissolved hydrophobic organic contaminants to microbial biofilms. Water research, 41, 2202-2210.

Wieder, K. J., King, K. R., Thompson, D. M., Zia, C., Yarmush, M. L., & Jayaraman, A. (2005). Optimization of reporter cells for expression profiling in a microfluidic device. Biomedical microdevices, 7(3), 213-222.

Wimpenny, J. (2000) An overview of biofilms as functional communities, p.1–24. In D. G. Allison, P. Gilbert, H. M. Lappin-Scott, and M. Wilson (ed.), Community structure and co-operation in biofilms. Cambridge University Press, Cambridge, United Kingdom.

Włóka, D. Kacprzak, M. Rosikoń, K. Fijałkowski, K., (2013) A study of migration of polycyclic aromatic hydrocarbons in a sewage sludge-soil system, Environment Protection Engineering, 39, 115-124.

Wolfaardt, G. M., Lawrence, J. R., Hendry, M. J., Robarts, R. D., & Caldwell, D. E. (1993). Development of steady-state diffusion gradients for the cultivation of degradative microbial consortia. Applied and environmental microbiology, 59, 2388-2396.

Wolfaardt, G. M., Lawrence, J. R., Robarts, R. D., Caldwell, D. E., (1995) Bioaccumulation of the Herbicide Diclofop in Extracellular Polymers and Its Utilization by a Biofilm Community during Starvation, Applied and Environmental Microbiology, 61, 152–158.

Wuana, R.A., Okieimen, F.E., (2011) HeavyMetals in Contaminated Soils: A Review of Sources, Chemistry, Risks and Best Available Strategies for Remediation, International Scholarly Research Network (2011), 1-20.

5 References

Wypych, G. (2004) Plasticizers use and selection for specific polymers (pp. 273-379). ChemTec Publishing: Toronto, Canada.

Yang, Z. Y., Greenstein, D., Zeng, E. Y., & Maruya, K. A. (2007). Determination of poly (dimethyl) siloxane–water partition coefficients for selected hydrophobic organic chemicals using ^{14}C-labeled analogs. Journal of Chromatography A, 1148, 23-30.

Yuan, S.Y., Chang, J.S., Yen, J.H., Chang, B.V., (2001) Biodegradation of phenanthrene in river sediment, Chemosphere, 43, 273–278.

Zhang, Y., Tao, S., (2009) Global atmospheric emission inventory of polycyclic aromatic hydrocarbons (PAHs) for 2004, Atmospheric Environment, 43, 812–819.

Zitomer, D.H., and Speece, R.E., (1993), Sequential Environments for Enhanced Biotransformation of Aqueous Contaminants, Environmental Science and Technology, 27, 227-244.

Zvezdov, A., & Zvezdova, D. (2010). A filtration water treatment device for colored waste water treatment. Annual Proceed.,"Angel Kanchev" University of Ruse, 49(9.1), 27-32.

6 APPENDIX

6.1 List of abbreviations

AAS	Atomic absorption spectrometry
BHJ	Bulk hetero conjunction
BTX	Benzene, toluene and xylene
DEHP	Di-2-ethylhexylphthalate
DIN	Deutsches Institut für Normung
EEC	European economic community
EEMF	Excitation emission matrix fluorescence
EEMF	Excitation emission matrix fluorescence
EI	Electron ionization
EPA	Environmental protection agency
EPDM	Ethylene propylene diene-monomerrubber
EPS	Extracellular polymeric substances
EQS	Environmental quality standards
FCS	Fluorescence correlations pectroscopy
FID	Flame ionization detection
FITC	Fluorescein isothiocyanate
FLD	Fluorimetric detection
FOG	Fats, oils and grease portion
FRAP	Fluorescence recovery after photo bleaching
GC	Gas chromatography

GIS	Geographic information system
HMW	High molecular weight
HOC	Hydrophobic organic compounds
HPLC	High-performance liquid chromatography
LC	Liquid chromatography
LC-DAD	Liquid chromatography with diode arraydetection
LD50	Letaldoseof50%
LLE	Liquid-liquid extractions
LMW	Low molecular weight
LOD	Limit of detection
LOQ	Limit of quantification
MS	Mass spectrometry detection
NIST	National institute of standards and technology
PAHs	Polycyclic aromatic hydrocarbons
PCB	Polychlorinated biphenyls
PCDD/PCDF	Polychlorinated dibenzodioxins/dibenzofuran
PDA	Photodiode array
PDMS	Polydimethylsiloxan
PM	Particulate matter
PN/PS	Polysaccharide/protein ratio
PRFS	Phase-resolved fluorescence spectroscopy
PTE	Potential toxic elements
QS	Quartz suprasil
SFQ	Selective fluorescence quenching

SFS	Synchronous fluorescence scan
SOM	Solid organic matter
SPE	Solid phase extractions
SPME	Solid phase micro extraction
SRB	Sulfate-reducing bacteria
TMV	Tobacco mosaic virus
TRF	Time-resolved fluorescence spectroscopy
USGS	United states geological survey
UV/Vis	Ultraviolet/visible
UVD	Ultraviolet detection
WFD	Water framework directive
WHO	World health organization
WRF	White rot fungi
WWTP	Wastewater treatment plant

ACKNOWLEDGEMENTS

I thank Prof. Dr. Hans-Curt Flemming for giving me the opportunity to be part of the ATWARM initiative (Advanced Technologies for Water Resource Management).

Thanks to Dr. Gerard Genuit and to all the members of his workgroup at the environmental administration (Umweltamt) of Bielefeld city. They provided me with several samples and walked me through a very interesting experience on the field.

Thanks to Dr. Ursula Telgheder and Dr. Klaus Kerpen, from the Department of Analytical Chemistry, for their scientific support. They have been mentors of inestimable value to me.

I am grateful to Dr. Andriy Kuklya, Robert Marks and Florian Uteschil, who were always willing to help me with my work. Together with Dr. Ursula Telgheder and Dr. Klaus Kerpen, they represented the core of an exciting and stimulating workgroup, in which I had the priviledge to discuss and realize many ideas.

I wish to thank Jan Frösler, from the Biofilm Centre. He helped me to unravel the many knots, which a foreign student might encounter on his way through a vast bureaucratic system such as the University of Duisburg-Essen. Among workshops and administration offices, we soon became known as "The duo". He always supported me during these three years offering to me his kind advice and invaluable friendship.

My great appreciation goes to the whole Duesseldorf Dragons Rugby Club, a group of friends who were always there when I needed them. In particular, my gratitude goes to John Thomson who helped me to refine my English writing skills.

I want to thank all my family for believing in me. In particular, my special thanks go to my parents, who taught me the importance of effort and responsibility at work.

I would like to express my utmost appreciation to Eva-Valeska for sustaining me with her invaluable love and patience. Her formidable strength, optimism and enthusiasm encouraged me in every moment. My gratitude for her goes beyond any possible word.

Printed by Books on Demand GmbH, Norderstedt / Germany